Methods in Molecular Biology

Series Editor
John M. Walker
School of Life and Medical Sciences
University of Hertfordshire
Hatfield, Hertfordshire, UK

For further volumes:
http://www.springer.com/series/7651

For over 35 years, biological scientists have come to rely on the research protocols and methodologies in the critically acclaimed *Methods in Molecular Biology* series. The series was the first to introduce the step-by-step protocols approach that has become the standard in all biomedical protocol publishing. Each protocol is provided in readily-reproducible step-by-step fashion, opening with an introductory overview, a list of the materials and reagents needed to complete the experiment, and followed by a detailed procedure that is supported with a helpful notes section offering tips and tricks of the trade as well as troubleshooting advice. These hallmark features were introduced by series editor Dr. John Walker and constitute the key ingredient in each and every volume of the *Methods in Molecular Biology* series. Tested and trusted, comprehensive and reliable, all protocols from the series are indexed in PubMed.

Microfluidics Diagnostics

Methods and Protocols

Edited by

Valérie Taly

Université Paris Cité, Centre de recherche des Cordeliers, INSERM, UMR-S1138, CNRS SNC 5096, Paris, France

Stéphanie Descroix

Institut Curie, UMR 168, Institut Pierre-Gilles de Gennes, Paris, France

Karla Perez-Toralla

Université Paris-Saclay, CEA, INRAE, Médicaments et Technologies pour la Santé (MTS), SPI, Gif-sur-Yvette, France

Humana Press

Editors
Valérie Taly
Université Paris Cité, Centre de recherche des Cordeliers, INSERM UMR-S1138, CNRS SNC 5096
Paris, France

Stéphanie Descroix
Institut Curie, UMR 168
Institut Pierre-Gilles de Gennes
Paris, France

Karla Perez-Toralla
Université Paris-Saclay, CEA, INRAE
Médicaments et Technologies pour la Santé (MTS), SPI
Gif-sur-Yvette, France

ISSN 1064-3745 ISSN 1940-6029 (electronic)
Methods in Molecular Biology
ISBN 978-1-0716-3849-1 ISBN 978-1-0716-3850-7 (eBook)
https://doi.org/10.1007/978-1-0716-3850-7

This Humana imprint is published by the registered company Springer Science+Business Media, LLC, part of Springer Nature.
The registered company address is: 1 New York Plaza, New York, NY 10004, U.S.A.

Paper in this product is recyclable.

Preface

For over 30 years, microfluidics has played a pivotal role in advancing innovative diagnostic approaches. In this new volume of *Methods in Molecular Biology*, we have selected a wide range of microfluidic-based approaches that exploit the unique features of microfluidic devices, holding significant potential in the field of diagnosis.

This work emphasizes how microfluidics can be harnessed to analyze a broad spectrum of biomarkers, spanning from small molecules like DNA to extracellular vesicles (EVs). Diagnosis, no longer confined to molecular bioanalysis, has evolved to single-cell analysis, which is a crucial application of microfluidics. This aspect could pave the way for next-generation diagnostics in the near future. Our exploration extends beyond the mere description of analysis processes. We also shed light on the crucial need to develop innovative approaches for sample preparation, a pivotal step in the analytical workflow. Furthermore, we investigate how microfluidic features facilitate the development of precise methods for analyzing bacteria and viruses, a field that gained substantial interest in direct correlation with global human health concerns. Lastly, we broaden the scope of diagnosis-based microdevices, examining their application in human cell-based culture, an emerging and promising facet of microfluidics. This selection of high potential development illustrates how microfluidics can participate in shaping the future of diagnosis.

Paris, France ***Valérie Taly***
Paris, France ***Stéphanie Descroix***
Gif-sur-Yvette, France ***Karla Perez-Toralla***

Contents

Part IV Microchips for Single Cell Analysis

Part V Microfluidic Chips for Analysis of Bacteria and Virus

Part VI Microfluidic Chips for Human Cell-Based Culture and Analysis

Contributors

LUCILE ALEXANDRE • *Institut Curie, Laboratoire Physico-Chimie (CNRS UMR 168), Institut Pierre-Gilles de Gennes, Sorbonne Université, PSL Research University, Paris, France*

MONICA ARAYA-FARIAS • *Université Paris-Saclay, CEA, INRAE, Médicaments et Technologies pour la Santé (MTS), SPI, Gif-sur-Yvette, France*

NATHAN AYMERICH • *Laboratoire de Colloïdes et Matériaux Divisés (LCMD), CBI, ESPCI Paris, Université PSL, CNRS, Paris, France*

AURÉLIEN BANCAUD • *CNRS, LAAS, Toulouse, France*

JEAN BAUDRY • *Laboratoire de Colloïdes et Matériaux Divisés (LCMD), CBI, ESPCI Paris, Université PSL, CNRS, Paris, France*

FRANÇOIS-CLÉMENT BIDARD • *Circulating Tumor Biomarkers Laboratory, SiRIC, Translational Research Department, Institut Curie, PSL Research University, Paris, France; Department of Medical Oncology, Institut Curie, Paris, France*

FRANÇOIS BOIZOT • *Univ. Grenoble Alpes, CEA, LETI, Grenoble, France*

AUDREY BOUTONNET • *ADELIS Technologies, R&D Department, Labège, France*

CAMILLE BRICTEUX • *CNRS, Institut Galien Paris-Saclay, Université Paris-Saclay, Orsay, France*

ARNAUD BRUNEEL • *INSERM UMR1193, Université Paris-Saclay, Orsay, France*

OLIVIA T. M. BUCHELI • *Laboratory for Functional Immune Repertoire Analysis, Institute of Pharmaceutical Sciences, D-CHAB, ETH Zürich, Zürich, Switzerland*

SOPHIE CAVASSILA • *Université Claude Bernard Lyon 1, INSA Lyon, UJM-Saint Etienne, CNRS, Inserm, CREATIS, Lyon, France*

SUNGHUN CHEONG • *Interdisciplinary Program in Bioengineering, Seoul National University, Seoul, South Korea*

HYERI CHOI • *Interdisciplinary Program in Bioengineering, Seoul National University, Seoul, South Korea*

VERONIQUE CHOTTEAU • *Department of Industrial Biotechnology, School of Engineering Sciences in Chemistry, Biotechnology and Health, KTH Royal Institute of Technology, Stockholm, Sweden*

FRANCESCO DAL DOSSO • *Department of Biosystems – Biosensors Group, KU Leuven, Leuven, Belgium*

ANNE-LAURE DEMAN • *Université Claude Bernard Lyon 1, CNRS, INSA Lyon, Ecole Centrale de Lyon, CPE Lyon, INL, Villeurbanne, France*

LUCIE DESCAMPS • *Université Claude Bernard Lyon 1, CNRS, INSA Lyon, Ecole Centrale de Lyon, CPE Lyon, INL, Villeurbanne, France; Université Claude Bernard Lyon 1, CNRS, Institut Lumière Matière, Villeurbanne, France*

STÉPHANIE DESCROIX • *Institut Curie, UMR 168, Institut Pierre-Gilles de Gennes, Paris, France*

FAHIMA DI FEDERICO • *Sorbonne Universités, UPMC Univ Paris 06, Paris, France; Institut Curie, PSL Research University, Paris, France*

PETRA S. DITTRICH • *Department of Biosystems Science and Engineering, ETH Zürich, Basel, Switzerland*

Simon Dumas • *Institut Curie, Laboratoire Physico-Chimie (CNRS UMR 168), Institut Pierre-Gilles de Gennes, Sorbonne Université, PSL Research University, Paris, France*
Klaus Eyer • *Laboratory for Functional Immune Repertoire Analysis, Institute of Pharmaceutical Sciences, D-CHAB, ETH Zürich, Zürich, Switzerland; Department of Biomedicine, Aarhus University, Aarhus C, Denmark*
Anna Fomitcheva-Khartchenko • *IBM Research Europe, Rüschlikon, Switzerland*
Marie Gaillard • *Univ. Grenoble Alpes, CEA, LETI, Grenoble, France; Univ. Grenoble Alpes, CNRS, CEA, Grenoble INP, IRIG, SyMMES, Grenoble, France*
Sonia Garrigou • *Université Sorbonne Paris Cité, INSERM UMRS 1147, CNRS SNC, Translational Research and Microfluidics, Equipe labellisée Ligue Nationale Contre le Cancer, Paris, France*
Frédéric Ginot • *ADELIS Technologies, R&D Department, Labège, France*
Saba Goodarzi • *Université de Lyon, Université Claude Bernard Lyon 1, CNRS, Institut Lumière Matière, Villeurbanne, France*
Noo Li Jeon • *Interdisciplinary Program in Bioengineering, Seoul National University, Seoul, South Korea; Department of Mechanical Engineering, Seoul National University, Seoul, South Korea; Institute of Advanced Machines and Design (SNU-IAMD) , Seoul National University, Seoul, South Korea*
Ailian Jin • *Interdisciplinary Program in Bioengineering, Seoul National University, Seoul, South Korea*
Petra Jusková • *Department of Biosystems Science and Engineering, ETH Zürich, Basel, Switzerland*
Govind V. Kaigala • *IBM Research Europe, Rüschlikon, Switzerland; School of Biomedical Engineering, Vancouver Prostate Center, Life Sciences Institute, University of British Columbia, Vancouver, BC, Canada*
André Kling • *Department of Biosystems Science and Engineering, ETH Zürich, Basel, Switzerland*
Jeroen Lammertyn • *Department of Biosystems – Biosensors Group, KU Leuven, Leuven, Belgium*
Emmanuelle Laurenceau • *Ecole Centrale de Lyon, CNRS, INSA Lyon, Université Claude Bernard Lyon 1, CPE Lyon, CNRS, INL, Ecully, France*
Damien Le Roy • *Université Claude Bernard Lyon 1, CNRS, Institut Lumière Matière, Villeurbanne, France*
Théo Liénard–Mayor • *CNRS, Institut Galien Paris-Saclay, Université Paris-Saclay, Orsay, France*
François Lux • *Université de Lyon, Université Claude Bernard Lyon 1, CNRS, Institut Lumière Matière, Villeurbanne, France; Institut Universitaire de France (IUF), Paris, France*
Thanh Duc Mai • *CNRS, Institut Galien Paris-Saclay, Université Paris-Saclay, Orsay, France*
Bongkot Ngamsom • *School of Natural Sciences, University of Hull, Hull, UK*
Henry Ordutowski • *Department of Biosystems – Biosensors Group, KU Leuven, Leuven, Belgium*
Valentina Palacio-Castañeda • *Department of Medical BioSciences, Radboud University Medical Center, Nijmegen, The Netherlands*
Nicole Pamme • *Department of Materials and Environmental Chemistry, Stockholm University, Stockholm, Sweden*

DOHYUN PARK • *Department of Mechanical Engineering, Seoul National University, Seoul, South Korea*

LÉA PAYEN • *Laboratoire de Biochimie et Biologie Moléculaire, CICLY UR3738, Groupe Hospitalier Sud, Hospices Civils de Lyon, Pierre Bénite, France*

IAGO PEREIRO • *Sorbonne Universités, UPMC Univ Paris 06, Paris, France; Institut Curie, PSL Research University, Paris, France; Institut Pierre-Gilles de Gennes for Microfluidics (IPGG), Paris, France; IBM Research Europe, Rüschlikon, Switzerland*

KARLA PEREZ-TORALLA • *Université Paris-Saclay, CEA, INRAE, Médicaments et Technologies pour la Santé (MTS), SPI, Gif-sur-Yvette, France*

INÊS F. PINTO • *KTH Royal Institute of Technology, Division of Nanobiotechnology, Department of Protein Science, Science for Life Laboratory, Solna, Sweden*

KEVIN PORTMANN • *Laboratory for Functional Immune Repertoire Analysis, Institute of Pharmaceutical Sciences, D-CHAB, ETH Zürich, Zürich, Switzerland*

CHARLOTTE PROUDHON • *Univ Rennes, Inserm, EHESP, Irset (Institut de recherche en santé, environnement et travail) – UMR_S 1085, Rennes, France*

CATHERINE PUDDA • *Univ. Grenoble Alpes, CEA, LETI, Grenoble, France*

MATHILDE RICHERD • *Institut Curie, Laboratoire Physico-Chimie (CNRS UMR 168), Institut Pierre-Gilles de Gennes, Sorbonne Université, PSL Research University, Paris, France*

CHARLOTTE RIVIÈRE • *Université de Lyon, Université Claude Bernard Lyon 1, CNRS, Institut Lumière Matière, Villeurbanne, France; Institut Universitaire de France (IUF), Paris, France; Institut Convergence PLAsCAN, Centre de Recherche en Cancérologie de Lyon, INSERM U1052-CNRS UMR5286, Centre Léon Bérard, Université de Lyon, Université Claude Bernard Lyon1, Lyon, France*

PABLO RODRIGUEZ-MATEOS • *Department of Materials and Environmental Chemistry, Stockholm University, Stockholm, Sweden*

YOANN ROUPIOZ • *Univ. Grenoble Alpes, CNRS, CEA, Grenoble INP, IRIG, SyMMES, Grenoble, France*

AMAN RUSSOM • *KTH Royal Institute of Technology, Division of Nanobiotechnology, Department of Protein Science, Science for Life Laboratory, Solna, Sweden; AIMES, Center for the Advancement of Integrated Medical and Engineering Sciences at Karolinska Institutet and KTH Royal Institute of Technology, Stockholm, Sweden*

NICOLAS SARRUT-RIO • *Univ. Grenoble Alpes, CEA, LETI, Grenoble, France*

STEVEN SCHMITT • *Myria Biosciences AG, Basel, Switzerland*

MARCO SERRA • *Institut Curie, Laboratoire Physico-Chimie (CNRS UMR 168), Institut Pierre-Gilles de Gennes, Sorbonne Université, PSL Research University, Paris, France*

DRAGANA SPASIC • *Department of Biosystems – Biosensors Group, KU Leuven, Leuven, Belgium*

VALÉRIE TALY • *Université Paris Cité, Centre de recherche des Cordeliers, INSERM, UMR-S1138, CNRS SNC 5096, Paris, France*

MYRIAM TAVERNA • *CNRS, Institut Galien Paris-Saclay, Université Paris-Saclay, Orsay, France*

AURÉLIE THUAIRE • *Univ. Grenoble Alpes, CEA, LETI, Grenoble, France*

INGA TIJUNELYTE • *CNRS, LAAS, Toulouse, France*

NGUYET-THUY TRAN • *CNRS, Institut Galien Paris-Saclay, Université Paris-Saclay, Orsay, France*

BAS VAN DE CROMMERT • *Department of Medical BioSciences, Radboud University Medical Center, Nijmegen, The Netherlands*

LORENZ VAN HILEGHEM • *Department of Biosystems – Biosensors Group, KU Leuven, Leuven, Belgium*
WOUTER P. R. VERDURMEN • *Department of Medical BioSciences, Radboud University Medical Center, Nijmegen, The Netherlands*
JEAN-LOUIS VIOVY • *Sorbonne Universités, UPMC Univ Paris 06, Paris, France; Institut Curie, PSL Research University, Paris, France; Institut Pierre-Gilles de Gennes for Microfluidics (IPGG), Paris, France*
DRIES VLOEMANS • *Department of Biosystems – Biosensors Group, KU Leuven, Leuven, Belgium*

Part I

Overview

Chapter 1

Self-Powered Microfluidics for Point-of-Care Solutions: From Sampling to Detection of Proteins and Nucleic Acids

Dries Vloemans,
Lorenz Van Hileghem,
Henry Ordutowski,
Francesco Dal Dosso, Dragana Spasic, and Jeroen Lammertyn

Abstract

Self-powered microfluidics presents a revolutionary approach to address the challenges of healthcare in decentralized and point-of-care settings where limited access to resources and infrastructure prevails or rapid clinical decision-making is critical. These microfluidic systems exploit physical and chemical phenomena, such as capillary forces and surface tension, to manipulate tiny volumes of fluids without the need for external power sources, making them cost-effective and highly portable. Recent technological advancements have demonstrated the ability to preprogram complex multistep liquid operations within the microfluidic circuit of these standalone systems, which enabled the integration of sensitive detection and readout principles. This chapter first addresses how the accessibility to in vitro diagnostics can be improved by shifting toward decentralized approaches like remote microsampling and point-of-care testing. Next, the crucial role of self-powered microfluidic technologies to enable this patient-centric healthcare transition is emphasized using various state-of-the-art examples, with a primary focus on applications related to biofluid collection and the detection of either proteins or nucleic acids. This chapter concludes with a summary of the main findings and our vision of the future perspectives in the field of self-powered microfluidic technologies and their use for in vitro diagnostics applications.

Key words In vitro diagnostics, Point-of-care testing, Remote microsampling, Self-powered microfluidics, Capillary forces

1 The Growing Role of Diagnostics in a Changing World

In the coming decades, the world population is expected to maintain its growth because of aging and increasing birth rates in developing countries. This, together with the overall societal shift toward a sedentary lifestyle and unhealthy dietary pattern,

---

Dries Vloemans, Lorenz Van Hileghem, Henry Ordutowski have equally contributed to this chapter.

Valérie Taly et al. (eds.), *Microfluidics Diagnostics: Methods and Protocols*, Methods in Molecular Biology, vol. 2804,
https://doi.org/10.1007/978-1-0716-3850-7_1,

accelerating rate of globalization, and along-going climate change effects, has been leading to an increasing prevalence of chronic and infectious diseases in both high-income and developing countries [1, 2]. In this context, in vitro diagnostics (IVD) plays a crucial role in global healthcare for identifying and monitoring diseases, as well as for providing patients with proper healthcare management (e.g., patient isolation in case of contagious infectious diseases or starting personalized therapy for those suffering from chronic diseases). Moreover, IVD provides an important tool for enabling early diagnosis and consequently a better medical outcome [3, 4]. This diagnosis is based on detecting cell abnormalities (e.g., cell count, size, and shape), tissue biopsies, or analytical (bio)markers in biofluids (e.g., blood and urine). The latter can comprise a variety of small inorganic chemicals (e.g., heavy metals, toxins, and electrolytes) or organic biochemicals (e.g., proteins, lipids, carbohydrates, and nucleic acids) compounds. As such, conventional IVD tests include cell culture, microscopy, immunology, and molecular-based methods, all of which are generally accepted and widely implemented as gold standards in clinical practice [5].

In developed (high-income) countries, most of these IVD tests are performed in centralized laboratories (i.e., hospitals or specialized test facilities) that are equipped with expensive semi- or fully automated high-quality devices, often capable of high-throughput testing (workflow illustrated in Fig. 1 for standard centralized blood test). Even though many tests can be run in parallel, a team of well-trained technical staff is necessary, combined with a proper logistic network (e.g., to transport samples from the collection site to the central laboratory). Moreover, a visit to the clinic and/or doctor's office is still necessary for specimen collection. The latter is often inconvenient and time-consuming, certainly for chronically ill patients who require regular monitoring of their disease status and treatment efficacy, resulting in reduced quality of life and increased healthcare costs.

In low- to middle-income countries (LMIC), accounting for the majority of world population, neither the financial resources, nor the infrastructure, are available for such an advanced

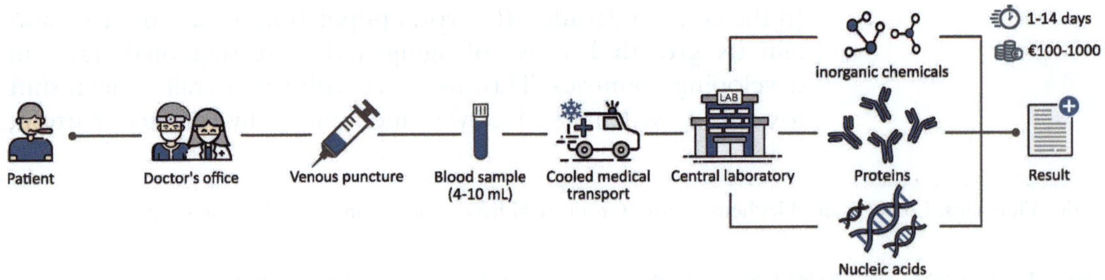

Fig. 1 Schematic workflow illustration for a standard blood test through a centralized healthcare strategy

centralized testing strategy [5–8]. The lack of accessibility to affordable and reliable diagnostic tests in developing countries is very contradictory to the high prevalence of, for instance, infectious diseases, which still account for millions of deaths in these regions on a yearly basis [7, 8]. People living in rural areas often must travel long distances to get tested, which constitutes a huge bottleneck in case of serious illness, lack of financial support, and stigma. Furthermore, because results are not immediately available, a high percentage of patients do not manage to retrieve their results and get proper follow up [9, 10]. This is problematic as early, and correct diagnosis is crucial for disease containment and administration of the appropriate treatment. Moreover, many patients receive empiric antimicrobial therapies without proper diagnosis, even in case of viral infections, which contributes to the occurrence of antimicrobial-resistant species [11, 12]. The growing problem of antimicrobial resistance and the inability to control recent disease outbreaks (i.e., COVID-19, Ebola, and Zika) [13–15] have raised a lot of concerns regarding the effectiveness of current centralized testing strategies and illustrated the need for decentralized accessible diagnostic tools [5, 16].

2 Increased Accessibility to Reliable Diagnostic Tools through Decentralized POC Solutions

To increase the accessibility to reliable diagnostic tests in resource-limited settings, defined here as any setting without direct access to specialized equipment and trained healthcare personnel (e.g., home, community center, doctors' office, airport, and school), worldwide healthcare systems have to shift more toward decentralized and patient-centric approaches [17]. To realize this, two different strategies can be implemented, with the workflows for a routine blood test illustrated in Fig. 2: (1) remote collection of sample specimens (like dried blood spot, DBS) and/or (2) bringing the diagnostic tests closer to the patient, so-called point-of-care

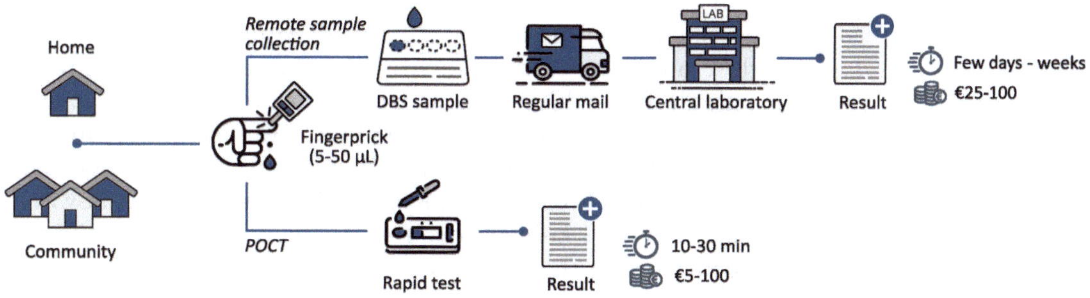

Fig. 2 Schematic illustration of two decentralized approaches consisting of remote sample specimen collection through DBS and POCT with rapid IVD tests (workflow presented for blood-based tests)

testing (POCT) [18]. Important to note is that, although both strategies are described here separately, they are complementary approaches with the preference determined by the urgency for clinical decision making and thus the need for an immediate diagnostic outcome (possibly giving up on quantitative information).

2.1 Remote Sample Collection

The recent advancements made in bioanalytical detection techniques have drastically reduced the required sample volume for accurate diagnosis and have become the main driver behind microsampling approaches [19]. Therefore, remote microsampling has gained a lot of interest and involves the collection of small volumes (<100 μL) of blood (e.g., finger-prick capillary blood in the form of DBS), saliva, or other biofluids in decentralized settings (e.g., at home), which are afterward shipped to a central laboratory for analysis. This microsampling procedure is minimally invasive and easy to perform as it does not require the presence of trained healthcare workers, even allowing for self-sampling. Moreover, the long-term stability of the dried sample circumvents the need for cooled transport [20, 21] and for the patients to travel long distances [22]. This can be beneficial for patients in developed countries suffering from chronic diseases or cancer (by replacing frequent doctor/hospital visits with remote self-sampling) but also in LMIC for routine screening (e.g., HIV viral load monitoring) and large-scale epidemiological studies (e.g., COVID-19 serosurveillance) of hard-to-reach populations [23–26]. In addition, microsampling techniques have also gained a lot of interest from the pharmaceutical industry to increase candidate enrolment and their retention in clinical studies for drug development, while simultaneously reducing trial budget [27]. Also, in preclinical studies, the use of microsampling is beneficial in terms of ethical considerations (i.e., animal welfare) and aligns well with the 3R strategy (reduce, refine, and replace), being widely adopted and implemented [28].

2.2 Near-Patient Testing through POCT

In some cases, rapid on-the-spot diagnosis is crucial to make quick clinical decisions in resource-limited settings. Therefore, the World Health Organization (WHO) called for the development of new diagnostic tools that can operate outside laboratory environments in close proximity to the patient with a fast turnaround time and that may lead to a direct change in patient management [8, 29–31], often referred to as POCT. The first POCT was introduced in 1962 by the development of a rapid test to monitor blood glucose levels at the bedside of hospitalized patients [32]. The principle, however, became only well-known after the commercialization of the first pregnancy test in the early 1980s [33]. To create a framework for developing the ideal test, which can be used at every level of the healthcare system in LMIC, the WHO published a set of criteria, earlier known under the ASSURED acronym (Table 1) [8, 34] and

Table 1
REASSURED criteria outlined by the WHO for POCT

Guidelines	Description	Specifications
Real-time connectivity	Tests are connected and/or a reader or smartphone is used to power the reaction and/or read test results to provide required data to decision makers	Embedded connectivity into the test or read-out instrument (e.g., smartphone) for centralized data collection of test results
Ease of specimen collection	Tests should be designed for simple use with noninvasive specimens	Compatible with raw unprocessed samples (capillary blood, urine, saliva, ...)
Affordable	Tests are affordable to end-users and the health system	Less than $500 for device Less than $10 per test
Sensitive	Minimal false negatives	Comparable to gold standard
Specific	Minimal false positives	Comparable to gold standard
User-friendly	Tests should be easy to perform with minimal user-intervention steps and require minimal user training with no prior knowledge of diagnostic testing.	Max 2–3 user-intervention steps and 1–2 days of training
Rapid	Results are rapidly available to ensure patient management and treatment during the same visit	Results within 15–60 min
Robust	Consistent performance, independent of applied setting (home, laboratory) and used supply chain (temperature, humidity, time delay, mechanical stress)	Shelf-life >1 year at room temperature
Equipment-free	Fully integrated without the need for complex equipment and infrastructure	Compact, battery-powered, on-site data analysis, easy disposal, easy sample handling
Environmentally friendly	Completed tests are easy to dispose	Manufactured from recyclable materials
Delivered to those who need them	Accessible to the end-user	Portable, hand-held

more recently as REASSURED (see further). The original acronym embodies three main principles: affordability, accuracy, and accessibility, among which trade-offs must be considered depending on the implemented level of the healthcare system [35]. For instance, highly sensitive and specific diagnostic tests (e.g., polymerase chain reaction (PCR)) require complex instrumentation and trained technicians for operation and are therefore restricted to a laboratory environment. As a consequence, these tests are not affordable and accessible to patients with limited available resources for healthcare. Chromatographic rapid diagnostic tests (RDT), on the other hand, are easy to use, low-cost and provide fast results independent of complex external instrumentation. Although these tests are widely accessible and satisfy many of the ASSURED criteria, they often lack in accuracy, resulting in false positive and negative results. Moreover, three decades of experience in implementing RDTs have revealed that many of the problems are not only originating from the test performance itself but are encountered during specimen collection and/or processing [36]. Together with the rapid advancements in digital technologies (internet of medical things [37], Industry 4.0 [38] and printed electronics [39]) along with environmental concerns, this has led in 2019 to the incorporation of three additional criteria to the list: "ease of specimen collection," "environmental friendly" and "real-time connectivity," giving birth to the REASSURED acronym (Table 1) [35].

3 Self-Powered Lab-on-a-Chip (LOC) Microfluidics

LOC technology has long promised to enable (advanced) IVD tests at the POC. This concept was first introduced in 1993 by Manz et al. as a miniaturized total chemical analysis system, demonstrating the manipulation of small volumes (nano- to microliter range) of liquids through microchannels for performing biochemical analysis [40]. In these devices, the unique fluid behavior at the microscale was exploited to speed up chemical reactions, while the small dimensions led to a drastic increase in throughput and parallelization capability with minimal reagent consumption [41, 42]. The advances made in manufacturing methods, originating from the microelectromechanical systems field, allowed for the fabrication and integration of several microfluidic valving and pumping mechanisms [43, 44]. These fluidic elements were key for automating complex liquid manipulations on microfluidic devices and formed the first steps toward integrated bioanalytical systems. The introduction of soft lithography-based rapid prototyping techniques, thereafter, enabled the production and replication of microfluidic systems in transparent elastomer polymers, such as polydimethylsiloxane (PDMS) within less than a day [45, 46]. This led to the widespread implementation of LOC devices and microfluidics in many scientific fields, including analytical chemistry, genomic and proteomic research, environmental monitoring, and the detection of biohazards [47].

The theoretical possibility to integrate and automate all laboratory procedures from sample to result within a single credit card-sized device by means of miniaturization makes this LOC technology the "holy grail" for advanced microsampling and POC diagnostic applications. Nevertheless, despite the numerous offered advantages, only a few LOC devices made it to the market [48]. This is because most LOC devices still lack integration by

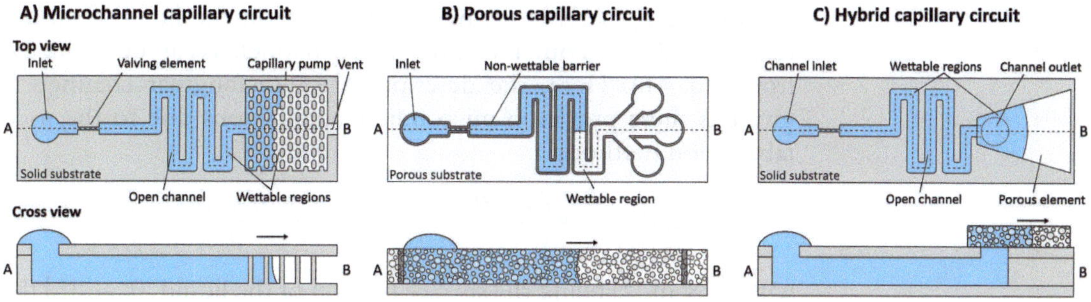

Fig. 3 Schematic illustrations of the different passive microfluidic pumping systems that exploit capillary forces in (**a**) wettable microchannels, (**b**) porous substrates, and (**c**) a hybrid combination of both to induce liquid flow. At the bottom, a cross-view of the microfluidic network is given along the direction of fluid propulsion (indicated by dotted line in the top view)

relying on external equipment for fluid handling (i.e., syringe pumps) and signal acquisition (i.e., microscopes), which require trained personnel for operation and maintenance. These ancillary devices are often expensive and require complex laboratory setups, which resulted in the mocking term "chip-in-a-lab" instead [49]. To make LOC devices more accessible, they must be decoupled from their laboratory environment. Hereto, self-powered microfluidics have emerged as an interesting alternative because the driving source for liquid manipulation is embedded within the device itself. Over the years, many self-driven microfluidic pumping strategies have been presented, employing a variety of passive driving sources, such as surface tension, gravity, vacuum, finger-actuation, osmosis, capillarity, and many more. Among these, capillary-driven microfluidic pumps have proven to be the most successful and widely implemented, fulfilling most of the REASSURED criteria for POC devices [35]. Therefore, the focus of this chapter is solely on capillary-based microfluidics, which can be divided based on the used matrix for fluid propulsion into microchannel, porous, and hybrid capillary circuits as conceptualized in Fig. 3. For an in-depth explanation of the other principles, the reader is referred to two extensive review papers [50, 51].

3.1 Microchannel-Based Capillary Microfluidics

In capillary microfluidics, the liquid is propelled through capillary action, which is the phenomenon of spontaneous liquid flow due to the interplay of cohesive and adhesive forces between the liquid and its solid surroundings. More specifically, when a liquid is confined within a microchannel, a capillary pressure (i.e., Laplace pressure) arises over its advancing liquid-air interface as a result of the surface tension and the curved liquid interface imposed by the surface properties (i.e., contact angle) of the surrounding channel walls [52–54]. In case of hydrophilic or wettable microchannels (water contact angle $<90°$), a negative capillary pressure is obtained, leading to capillary filling of the microchannel. This capillary pressure (ΔP_c [Pa]) is strongly dependent on the liquid's surface tension and the geometry of the microchannel, as can be deducted from Eq. (1). Here, the description for rectangular channels is given as these prevail in microfluidic applications due to the used fabrication methods:

$$\Delta P_c = -\gamma\left(\frac{\cos\theta_t + \cos\theta_b}{h} + \frac{\cos\theta_l + \cos\theta_r}{w}\right), \tag{1}$$

with γ [N·m^{-1}] being the surface tension of the liquid, $\theta_{t,b,l,r}$ [°] the contact angle of the corresponding top, bottom, left, and right channel walls, and h, w [m] the height and width of the rectangular microchannel, respectively. Equation 1 shows that capillary action only becomes significant at the microscale (w and h < 1 mm) where surface effects start to dominate over bulk forces (e.g., gravity). By adjusting the geometry of the cross-section (w × h) and thus the local capillary pressure within capillary circuits, the fluid flow can be controlled very precisely. This capillary flow, Q [m^3·s^{-1}], can be described by Eq. (2), in which ΔP [Pa] resembles the capillary pressure difference across the microchannel, μ [Pa·s] the viscosity of the liquid, and $L(t)$ [m] relates the position of liquid meniscus within a rectangular microchannel (with h < < w) in function of time t [s] [55, 56]:

$$Q = \frac{h^3 w \Delta P}{12\mu L(t)}\left[1 - 0.630\frac{h}{w}\right] \tag{2}$$

The use of capillary microfluidics in the biomedical field was first exploited in 1997 by Delamarche and colleagues [57], followed by the development of several capillaric elements to enable complex fluid operations in standalone capillary circuits (Fig. 4a) [55, 58, 59]. Conventional capillary systems were subjected to a limited pump capacity and nonconstant flow rates (Eq. (2)) because the capillary forces only occur at the advancing liquid-air interface within the microchannels. Consequently, the liquid flow stops upon complete filling of the channel network and it becomes hard to control the liquid flow. To resolve this, micro-structured capillary pumps were introduced at the end of the capillary circuits to generate additional controlled capillary pressure (suction) [60–62]. By selecting the proper design and geometry of the microstructures within the pump, the operation time, volume, and flow rate of drawn-in liquid can be precisely regulated (Fig. 4b) [63]. Moreover, to achieve controlled and coordinated fluid manipulation, several capillary valving principles (retention, stop, burst, and trigger valves) have been developed. These valves use abrupt 2- or 3D geometry changes [58, 64, 65] or alterations in surface chemistry (i.e., contact angles) [66, 67] to locally adjust the capillary pressure, and thus capillary flow, within the microchannel (Fig. 4c–e). The integration of 1 or multiple of these valves allows

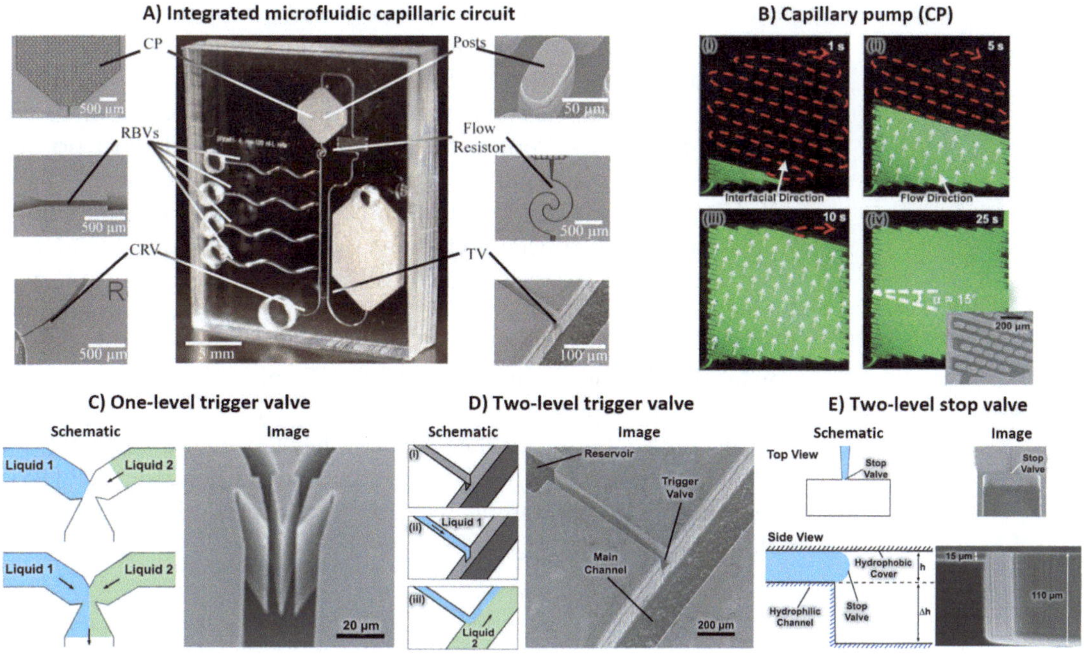

Fig. 4 (**a**) Example of a standalone PDMS-based microfluidic capillary circuit integrating multiple capillaric elements for sequential delivery of four reagents to a reaction chamber. Abbreviations used: *CP* capillary pump, *RBV* retention burst valve, *CRV* capillary retention valve, and *TV* trigger valve. (**b**) Liquid filling behavior in a serpentine-shaped capillary pump in which the flow direction is guided by an array of microposts introducing zones with high and low capillary pressure. Schematic working principle and scanning electron microscopy image of different capillary elements to stop liquid flow using abrupt 2D or 3D expansions in the microfluidic network: (**c**) One-level trigger valve, (**d**) two-level trigger, and (**e**) two-level stop valve. Reproduced with permission from (**a**) Safavieh and Juncker. ©2013 Royal Society of Chemistry; (**b**) Safavieh, Tamayol, and Juncker. ©2015 Springer-Verlag; (**c**–**e**) Olanrewaju, Beaugrand, Yafia and Juncker. ©2018 Royal Society of Chemistry

for complex liquid handlings on capillary circuits such as time delays as well as incubation, mixing, and even sequential delivery of reagents, showing its potential for POC assays. Traditionally, complex and expensive cleanroom-based microfabrication techniques were required to introduce capillary channels in hydrophilic glass and silicon materials [68]. These have been replaced by soft lithography enabling fast and controlled microchannel fabrication in polymer materials. However, a drawback for the latter is that often an additional hydrophilic treatment step is required to induce capillary flow within the microfluidic network.

3.2 Capillary Flow in Porous Matrices

The spontaneous liquid absorption properties within a variety of porous substrates such as paper [69–71], wool [72], cotton yarn [73, 74], or polyester have been exploited for flow control as well. Like for capillary microchannels, the fluid transport is also driven by the interplay between cohesive and adhesive forces inside the pores of the medium. The latter is being dictated by the mechanical (i.e., pore size, density, and distribution) and chemical (i.e., surface energy) properties of the porous material, which determine the liquid-wicking behavior [75]. Different models were built to describe the fluid transport within these porous media. In classic models, the Lucas–Washburn equation was extended to porous media, which were assumed as a parallel bundle of capillary channels [76, 77]. This flow description is derived under many assumptions, such as a constant cross-section, uniform pore size, and distribution, no external effects of gravity and evaporation, and hence, is an oversimplification. In reality, most porous materials, like paper, have a complex mechanical structure as they often consist of a random network of overlapping fibers. Therefore, more general approaches for describing the fluid transport in porous media based on Darcy's law have been developed [78–83]. This law (Eq. (3)) relates the pressure difference (ΔP [Pa]) over the length of a fully wetted porous medium to a certain volumetric flow rate ($Q\,[\mathrm{m}^3{\cdot}\mathrm{s}^{-1}]$) [78, 79].

$$Q = -\frac{\kappa A}{\mu L}\Delta P \tag{3}$$

Here, κ [m^2] resembles the permeability of the porous medium, which is a material property that measures the ability of a porous medium to transmit fluids and strongly depends on the pore density and effective pore radius. A [m^2], instead, defines the cross-sectional area of the porous matrix that is perpendicular to the flow.

The use of paper as a liquid transport medium offers many advantages as it is a widely available and inexpensive material, biocompatible, easily disposable, chemically and mechanically easily modified (e.g., coating, compressing, cutting, stacking, folding), enables efficient biomolecule functionalization (high surface area per volume), and compatible with mass-scale production processes (i.e., inkjet printing and roll-to-roll manufacturing) [84]. These are the main reasons for the success of chromatographic lateral flow assays (LFAs). However, a drawback is that they only enable one-directional flow, limiting their use for complex analytical assays. To resolve this, the group of Whitesides proposed the use of photolithography to pattern small millimeter-sized channels in paper substrates [71]. By defining microfluidic networks with hydrophobic boundaries, liquids could be manipulated in 2D, resulting in so-called micro paper-based analytical devices (μPAD). Based on this pioneering work, several other fabrication methods (e.g., wax printing [85, 86], plasma treatment [87], inkjet etching [88], and plotting [89, 90]) were presented, which facilitated the manufacturing process and led to an explosive growth of the field. To allow flows in both the vertical and horizontal direction and consequently improve their flexibility, 3D μPADs were developed by stacking or folding multiple 2D planar microfluidic

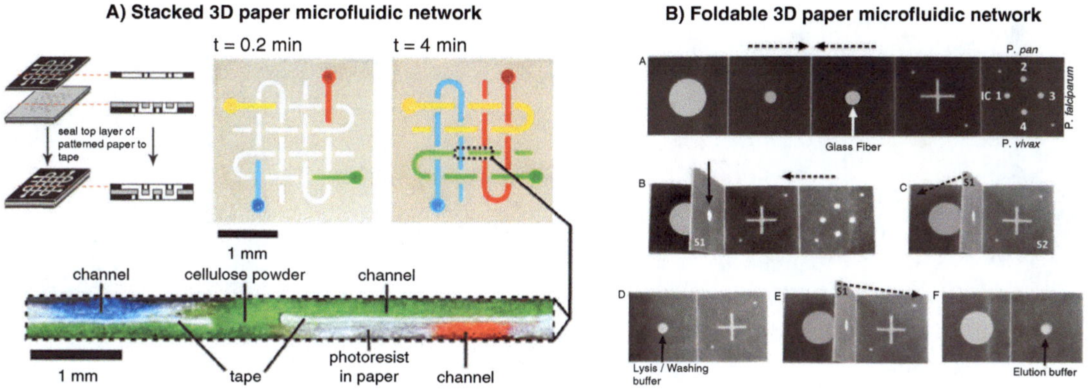

Fig. 5 Representative examples of 3D μPADs fabricated by: (**a**) stacking multiple layers of adhesive tape and patterned papers or (**b**) using origami folding of different patterned paper-fluidic networks to form a 3D fluidic network. Adapted with permission from (**a**) Martinez et al. ©2008 The National Academy of Science of the USA; (**b**) Xu et al. ©2016 Wiley

networks on top of each other (Fig. 5) [69, 70, 91, 92]. The simplicity to modify the design (i.e., channel length and width) and local composition (i.e., dissolvable sugar content, hydrophobic wax, compressed pores) of the paper network with modern patterning techniques led to the development of many functional fluidic elements for valving (i.e., delay valves [93, 94] and diode valves [95]), reagent splitting and mixing [70, 92], and the separation of plasma from whole blood [96, 97]. By incorporating this know-how within the design of paper networks, flow control was improved. This expanded the capability of μPADs to execute multistep sequential fluidic operations (i.e., sequential reagent delivery) and subsequentially multistep immunoassays [98–100].

However, a major issue with paper-based microfluidics remains the inherent variability of material properties such as pore size and porosity, which still hamper precise flow control. To resolve this, alternative fabrication methods such as 3D printing or photo structuring are being explored to create artificial porous substrates in a more reproducible and controlled fashion [101, 102]. Paper microfluidics have also difficulties to transport cells or other large particles (μm range), while unwanted interactions (e.g., cross-reactivity, nonspecific binding) between sample matrix compounds and the porous substrate can lead to a decreased effectiveness of assays [103]. Additionally, large sample and reagent volumes are often required, while reagent storage has proven to be difficult, certainly in the uncontrolled ambient environments at POC settings [104, 105].

3.3 Hybrid Capillary Microfluidics

To combine the best of both worlds, hybrid capillary microfluidic systems have been proposed, in which closed microchannels are used to take advantage of reduced sample-matrix interferences and

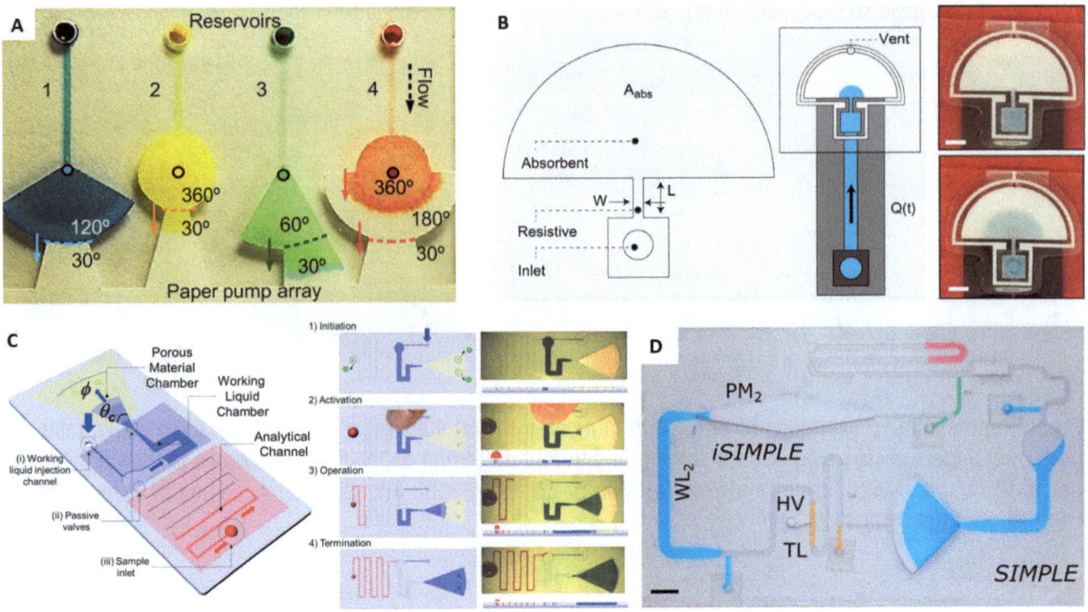

Fig. 6 Representative examples of porous-channel hybrid microfluidic capillary circuits. (**a**) Generation of different flow profiles in a microfluidic channel by using paper fans with two-phase and three-phase varying angles. (**b**) Schematic of a single hydraulic battery as a separate unit and connected to a microchannel. Pictures showing the fluid absorption in the hydraulic battery at different time points. (**c**) Configuration of the SIMPLE pump with schematic and pictures of the different steps of fluid behavior upon finger-press activation. (**d**) Picture of Integrated (i)SIMPLE chip combining the SIMPLE and iSIMPLE pump modes for shuttling of a sample liquid (red) back and forward. Abbreviations used: *PM* porous material, *WL* working liquid, *HV* hydrophobic vent, and *TL* trigger liquid. Reproduced with permission from (**a**) Wang, Hagen, and Papautsky. ©2013 American Institute of Physics; (**b**) © 2017 Cummins, Chinthapatla, Lenin, Ligler, and Walker; (**c**) Kokalj, Park, Vencelj, Jenko, and Lee. ©2014 The Royal Society of Chemistry; (**d**) Dal Dosso, Tripodi, Spasic, Kokalj and Lammertyn. ©2019 American Chemical Society

lower sample-volume requirements, whereas porous materials increase the liquid absorption capacity and thus pumping duration. Wang and colleagues, for instance, connected a fan-shaped paper substrate to the outlet of conventional polymer-based microchannels to generate a steady liquid flow by means of capillary action [106]. By adjusting the angle and size of the paper, a wide range of volumes (20–40 μL) and flow rates (20–100 μL/min) were achieved. Furthermore, they showed that flow rates can be programmed during the pumping process by connecting multiple pumping elements in series with varying fan angles along its radius (Fig. 6a). Similar concepts have been presented by others to enable sample delivery for DNA and multiplex protein detection assays [107, 108]. To improve flow consistency, Cummins and colleagues introduced a fluid resistance between the paper inlet and absorbent regions of their so-called hydraulic batteries, showing that by tuning the width–length ratio of this paper restriction, highly controlled flow rates can be achieved (Fig. 6b) [109]. Stacking multiple of these pumps with different geometries also enabled to generate complex flow profiles (e.g., step-down and linearly decreasing flow rate profiles). Even more complex profiles (i.e., oscillating) were obtained by separating the different layers with dissolvable films, acting as fluidic time delays.

Although these approaches drastically increase the pumping capacity and duration, it only allows for one-directional flow, thereby limiting the ability to perform complex fluid operations. Moreover, the sample liquid has to be in direct contact with the porous material, which makes the flow rate dependent on the sample fluid properties (i.e., viscosity and surface tension). These are known to vary strongly for different biofluids (e.g., blood vs. saliva) and in between patients (e.g., due to varying blood hematocrit (Hct) levels). To overcome these problems, the so-called (i) SIMPLE ((infusion) self-powered microfluidic pump by liquid manipulation) technology was proposed as a promising alternative [110, 111]. In this pumping concept, the spontaneous wicking action of a working liquid (e.g., water) into a precut porous material (e.g., filter paper), after a finger-press actuation, is exploited as a passive driving source to generate pressure gradients within the connected microfluidic channel network (Fig. 6c). Two different pump modes were developed that, depending on the used configuration, enable the pulling (i.e., SIMPLE pump) or pushing (i.e., iSIMPLE pump) of liquids. The ability to combine both pushing and pulling pumping modules together with the recently developed valving tools (i.e., hydrophobic burst and stop valves) has led to the establishment of a microfluidic toolbox, which is rapidly expanding and enables the performance of complex liquid operations such as flow splitting, plug shuttling, and volume metering (Fig. 6d) [112, 113]. This has been translated into multiple integrated systems applied to various fields such as microsampling [112], IVD [114], and (vaccine) delivery [111]. A big advantage of this system is that the sample flow is fully driven by the separated (i)SIMPLE pumping mechanism and therefore does not require hydrophilization of the channel network. Consequently, hydrophobic channels can be used for improving flow control, as described by McNeely and colleagues [115]. Furthermore, the flow rate is only dependent on the fluid properties of the working liquid, which remains unchanged for all chips.

4 Microfluidic Solutions for Remote Sampling

Although the microfluidic principles described in the previous section mainly focus on simplifying and integrating lab-based protein and nucleic acid detection assays in standalone platforms, an often overlooked challenge in developing these POC systems is the collection and application of the sample to them. The latter is, however, critical because the accuracy of the test is directly dependent on the quality of the applied sample. Venous blood is until present the golden standard sample type in laboratory testing for the majority of analyses, collected by the invasive procedure of phlebotomy using hypodermic needles. This technique requires well-trained healthcare personnel working according to good medical practices. However, decentralization demands the minimally invasive collection of readily available samples in a user-friendly way. To accomplish this, capillary blood and interstitial fluid (ISF) are being looked at as primary alternatives to venous blood as (1) they are both accessible from the skin's superficial layers, (2) they have a rich proteome, genome, and metabolome, thereby offering a possibility to detect disease-specific analytes as early as possible even with low-abundant proteins and nucleic acids, and (3) major steps have been already taken in their high-quality collection at the POC. Although other body fluids, such as saliva, sweat, urine, and tears, have been gaining interest lately [116–118], some challenges remain, including complex biochemical and physical fluid properties, unknown analyte stability, and low or highly variable concentrations, which necessitates application-specific research before they can really impact the medical world.

4.1 Capillary Blood Collection

Capillary blood can easily be reached by a finger- or heel-prick using a lancet [119]. This way, multiple capillaries in the dermal layer, which has a density of 14–30 capillary loops per mm^2 [120], are ruptured, releasing the blood (see also Fig. 7a). Its reduced

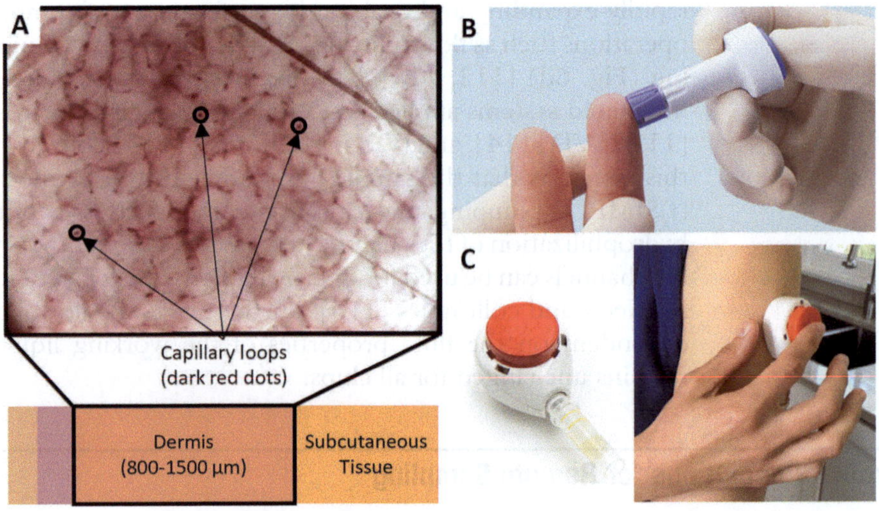

Fig. 7 Extraction of capillary blood. (**a**) Capillaries network in the volar forearm obtained by video-capillaroscopy showing perpendicularly arranged loops as dark red dots. (magnification ×100). (**b**) Disposable 23 G finger-prick lancet, which can be set in three different lengths. (**c**) Integrated device named Tasso+ for standardized capillary blood collection using the lancet poke-and-release principle. Reproduced with permission from (**a**) Humbert et al. ©2016 Springer Cham.; (**b**) ©2018 Roche Diabetes care. (**c**) ©2023 Tasso, Inc.

invasiveness makes this method suitable for self-sampling, due to which it became the primary method for at-home self-testing blood glucose levels since the 1980s. Nowadays, finger-pricking (Fig. 7b) is gaining interest for other applications requiring regular blood checks, like therapeutic drug monitoring (TDM), pharmacokinetic studies, or population health screening [121–123]. Furthermore, it is a possible solution to counteract the current issue of needle reuse and lack of medical trained personnel in the developing world (e.g., for HIV viral load monitoring) [124, 125].

Next to reduced invasiveness, the collected volumes via finger-prick sampling are significantly reduced in the range of tens of microliters, which is often more appropriate for analysis compared to the several milliliters sampled by venous puncture. However, adequate volumes can often only be achieved by squeezing the finger-tips [126, 127], which increases the risk of hemolysis and coagulation and might dilute the sample with intra- and intercellular fluid biasing the analytical results [128, 129]. Besides, operator errors including insufficient hand washing or inadequate blood application still remain the largest source of inaccuracies [130, 131]. To standardize capillary blood collection, the company Tasso developed an integrated device for the collection of 200–600 μL of capillary blood (Fig. 7c). The device has to be applied on the upper arm and upon one finger push, a lancet is released and retracted, while blood flows into a collection tube by the generation of a vacuum. Depending on the envisioned application, different collection tubes are available, containing different or no anti-coagulants.

4.2 Microfluidic Systems for DBS Sampling

If direct analysis is not intended, the collected blood sample should immediately be stabilized to maintain its quality during storage or transport to the analytical laboratory. Because liquid storage in collection tubes is still prone to deterioration, DBS offers a valuable alternative by drying the sample on absorbent filter paper. Today, DBS sampling is gaining clinical importance for its increased analyte stability during transport, logistical convenience and overall reduced cost [132]. Furthermore, regarding centralized laboratories, the method allows for automated sample preparation and subsequent analysis. Typically, one or multiple sub-punches are taken from the DBS and transferred into an extraction solvent to reconstitute the analytes.

Despite the advantages, some critical limitations detain DBS sampling from reaching its full potential. First, operator errors often result in inadequate DBS samples by e.g., spotting insufficient volumes, depositing the blood in the wrong location or overlapping spots, sending insufficiently dried samples, or contaminating after direct finger contact. Second, the high inter- and intra-person variation in blood Hct levels (i.e., volumetric percentage of red blood cells) causes analytical bias as a result of the direct proportionality to its viscosity. The higher the Hct level, the lower the diffusion through the filter paper leading to smaller spot areas. Hence, these sub-punches contain proportionally more analyte compared to the DBS sub-punch with a low Hct level, which is called the Hct-based sampling bias.

To solve this Hct bias, in 2014, Denniff and Spooner introduced the concept of volumetric sampling to replace the current practice of sub-punching [133]. Their sampler, branded as Volumetric Absorptive Micro Sampler (VAMS®), consists of a plastic handle and a hydrophilic porous tip to absorb precisely 10, 20, or 30 μL of blood. In essence, this tip is optimized to avoid Hct bias upon extraction as a whole. Studies comparing VAMS® against traditional DBS [134] and pipetted reference [135] show high reproducibility with CV-values between 3.9% and 4.3%. However, patient self-sampling validation revealed that VAMS® is still prone to user-dependent variability by over- or undersampling, contamination by direct finger contact, or the blood drop touching the sample holder [136]. Hence, obtaining high-quality samples using VAMS® can only be assured when following the instructions carefully.

User-dependency can be minimized by physically decoupling sample application and metering from DBS loading, using passive microfluidics. This also provides a visible proof of both successful sample metering and DBS loading by checking if the metering channel has been completely filled and emptied, respectively. Hereto, HemaXis DBS10 (Fig. 8a) combines a self-filling capillary metering channel with a traditional DBS card [137]. Upon loading a 10 μL blood sample, the blood drop after finger-pricking has to be brought in contact with the capillary inlet. After completely filling the capillary, closing the cartridge brings the filter paper into contact with its outlet to generate the DBS. By repeating this process, up to 4 DBS samples can be generated per cartridge. Capitainer®B (Fig. 8b) is another example using a 13.4 μL (and 50 μL with the Capitainer®B 50) self-filling capillary channel but loads automatically a pre-cut DBS, further avoiding manual handling steps [138]. A blood drop of minimally 18 μL has to be applied at once on the sample inlet hole with hydrophobic surroundings, upon which the excess blood is removed from the inlet before loading the DBS. This is achieved by the timely disintegration of a dissolvable polyvinyl alcohol film on top of the waste and DBS filter papers. The disadvantage of both HemaXis DBS10 and Capitainer®qDBS is that multiple sample applications are required for generating multiple DBS on a single cartridge. To avoid the need for multiple finger-pricking, HemaPEN® (Fig. 8c) meters four samples from the same drop in parallel in a pen format [139]. However, designed sample volumes are only 2.7 μL, limiting their applicability. Furthermore, the device is still prone to user mistakes. Indeed, the pen tip has to be positioned correctly and adequately

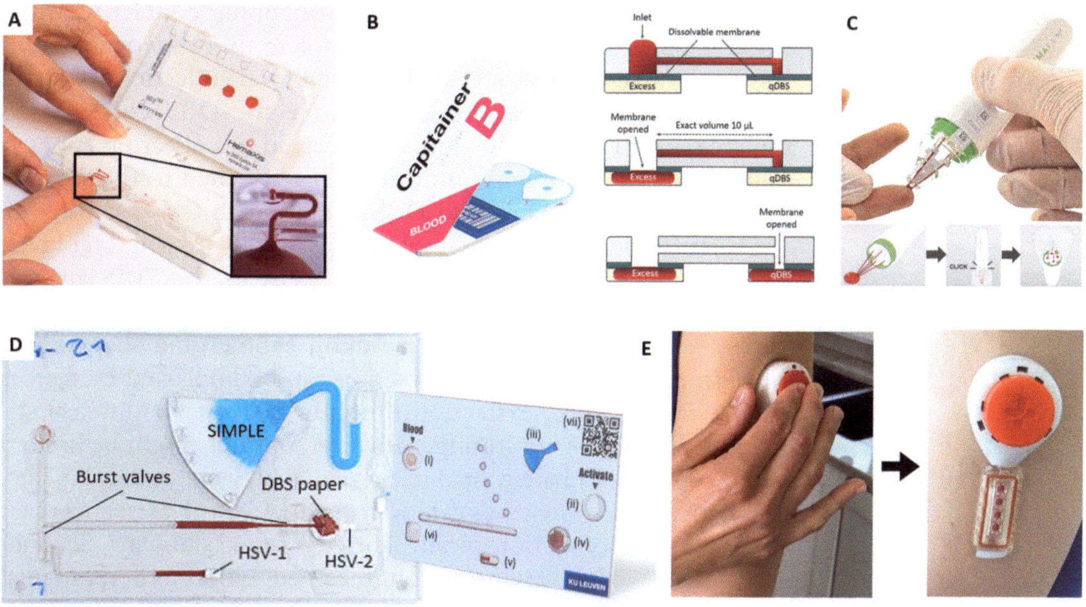

Fig. 8 Microfluidic solutions for volumetric DBS sampling. (**a**) HemaXis DBS10 generating up to four metered DBS by loading blood samples in the self-filling capillaries and closing the cartridge. (**b**) Capitainer®qDBS using a self-filling capillary channel in combination with a precut DBS and automated sample excess removal. (**c**) HemaPEN® allowing for parallel sampling of four DBS in a pen format. (**d**) Volumetric DBS actuated by SIMPLE. Precisely metered plug of 10 μL is split off from unknown sample volume by burst valves and hydrophobic stop valves (HSV) (**e**) Tasso M-20 integrating capillary blood collection with volumetric DBS sampling upon single press on a button. Reproduced with permission from (**a**) ©2018 Hemaxis; (**b**) ©2021 Capitainer; (**c**) ©2020 Trajan Scientific and Medical; (**d**) Vloemans et al. ©2022 Springer; (**e**) ©2023 Tasso, Inc.

long into the blood droplet, while DBS loading is based on the gravitational force by turning the pen upside down.

Alternatively, the need for fully automated parallel sample metering and further propulsion was achieved by sequential retention-and-release actions in microfluidic channels by hydrophobic burst valves with different burst pressures (Fig. 8d). This concept allows for metering multiple volumes between 5 and 15 μL in parallel, starting from a single sample drop. By the implementation of this metering concept with SIMPLE technology, a proof-of-concept volumetric DBS device was analytically validated in the context of TDM on a single, pre-cut DBS [112]. Moreover, the unique combination of channel-based microfluidic blood processing and paper-based microfluidic pumping also allows to implement multiple DBS loading in parallel or other processes downstream on the chip.

Although the discussed microfluidic solutions for DBS sampling (semi-)automate sample metering and subsequent DBS loading, sample collection is still done by finger-pricking. This potentially leads to problems like, for instance, failing to drop the capillary blood into the inlet or not maintaining contact with the blood droplet as long as needed. Complete user-independency is only achievable when all steps, including blood collection itself, are integrated. Today, only one such device for DBS collection is available, i.e., Tasso M-20 (Fig. 8e). Instead of collection tubes, a volumetric DBS sampling unit can also be attached to the sampler to load four precisely metered DBS samples with CV of <5%.

Next to the Hct-based sampling bias, highly varying Hct levels also have an effect on the following steps in the DBS analysis workflow. First, the remainders of the red blood cells can for some analytes hamper the release upon submerging the DBS in the extraction solvent. This extraction can be improved by factors such as temperature, extraction time, the solvent, or physical stimuli like shaking or sonicating. However, options might be limited in the context of protein or nucleic acid analysis to avoid damage. Alternatively, in some cases can be corrected for the extraction bias by the addition of an isotopically labeled internal standard. Second, the presence of a varying amount of red blood cell remnants, e.g. hemoglobin, in the matrix also leads to an altered response by interfering with the signal detection or the assay elements. In case whole blood samples are not essential, upstream plasma separation and, hence, loading dried plasma spots can circumvent Hct-based issues. Although microfluidic solutions for further sample processing are currently limited, membrane-based plasmapheresis has been used by Hauser et al. and Telimmune™ (previously Noviplex™) [140, 141].

4.3 Microneedle-Based Sample Collection

Despite reduced invasiveness, lancets still trigger significant pain perception and needle phobia. This holds the risk of losing compliance with regular health monitoring [142]. Also, regular use of lancets holds the risk of severely disturbing microcirculation of the skin. As a solution, microneedles have been gaining a lot of interest in the latest decades by further reducing invasiveness, being easier for use, and getting much higher patient acceptance [143]. Microneedles are needle-like projections, either configured in-plane or out-of-plane, with a diameter in the micrometer range and can be either solid or hollow. Solid microneedles are mostly used in a poke-and-release methodology, similar to lacing devices. YourBio Health's TAP II device (Fig. 9a) has an integrated array of solid microneedles to puncture skin capillaries. Upon the creation of a vacuum, the released blood is drawn into a microfluidic channel, mixed with anti-coagulants, and collected in a tube that can be sent back to the analytical laboratory [144]. Hollow microneedles, on the other hand, create a transdermal duct through which the released sample can be withdrawn, referred to as a poke-and-flow approach. By inserting a wick, reducing the lumen diameter, or making the material's surface hydrophilic, capillarity can be used to passively retract the released body fluid [145–147]. However, in

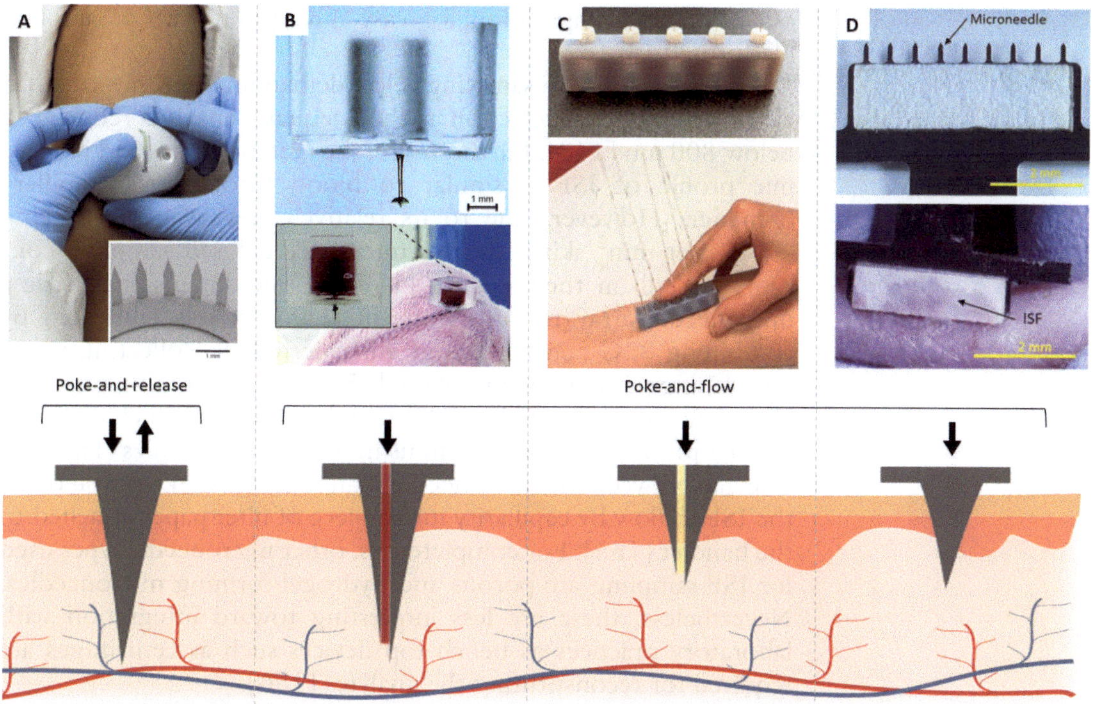

Fig. 9 Microneedle-based sample collection from skin. (**a**) TAP II device in a poke-and-release approach using long solid microneedles for capillary blood collection. (**b**) Single hollow microneedle-based proof-of-concept drawing capillary blood under prestored vacuum in a poke-and-flow approach. (**c**) Short hollow microneedles combined as an array to sample ISF upon capillary filling of the lumen. (**d**) Short solid microneedles combined with a filter paper patch for the collection of 2 μL ISF within 1 minute. Reproduced with permission from (**a**) Blicharz et al. ©2018 Nature; (**b**) Li et al. ©2015 Royal Society of Chemistry; (**c**) Miller et al. ©2018 Nature; and (**d**) Kolluru et al. ©2019 Springer

general, sample volumes obtained by capillary suction are small and in the range of a few microliters per needle [148]. This is regularly solved by combining multiple microneedles in arrays and/or creating an external vacuum. Hereto, Li and colleagues used a prestored vacuum inside a PDMS chamber to integrate a negative pressure gradient source with a single microneedle (Fig. 9b) [149]. The latter was fabricated by lithography, nickel electroplating, and laser cutting.

Hollow microneedles have an important advantage over solid microneedles as the latter need a mechanical system to release again after punching. However, hollow microneedles are more challenging to fabricate by a scalable production technology without jeopardizing their strength to avoid rupture or buckling upon insertion. Examples using small-scale 3D-printing or micro-injection molding currently still need an extra micromachining step to make the fluidic connection through the base plane [150, 151]. Furthermore, shaft lengths above 1500 μm are required to reach capillary blood vessels in the skin, which is even more challenging to achieve without losing aspect ratio and/or out-of-plane configuration for easy integration with a sampling platform [148, 152].

Alternatively, ISF sampling is less demanding in terms of fabrication and material selection as the adequate length is reduced to below 800 μm [148]. Furthermore, the metabolomics and proteomic profile of ISF is similar to blood through transcapillary exchange. However, due to its relatively low availability being <1 μL per mm^2 skin, effective microneedle-based sampling concepts are still in their infancy [153]. A one-dimensional hollow microneedle array proof-of-concept (Fig. 9c) was fabricated by combining 32G pen needles in a plastic holder, collecting up to 16 μL after 1–2 h of sampling [154]. However, this sampling efficiency was improved by Kolluru and colleagues, which collected up to 2 μL of ISF within 1 min using solid microneedles (Fig. 9d). The gap between the microneedle shafts and the skin tissue allowed the ISF to flow by capillarity into a piece of filter paper attached to the handler [155]. For completeness, other microneedle types used for ISF sampling are porous and hydrogel-forming microneedles. Nevertheless, these are less interesting toward integration with laboratory practices as bench-top devices such as centrifuges are required for reconstituting the analytes [156].

4.4 Toward Sample In–Result Out Systems

If rapid answers are essential, the full integration of hollow microneedle (arrays) with a POCT microfluidic device is key, thus providing sample collection, preparation, and analysis [156]. However, integrated solutions without the need for external syringe pumps or manual handling steps barely exist nowadays. A few examples that combine microneedles with paper-based biosensors by direct contact are discussed. Lee and co-workers connected porous PDMS microneedles for sample collection from a skin-mimicking agarose gel to filter paper for glucose detection by an enzymatic colorimetric reaction (Fig. 10a) [157]. Porous PDMS microneedles were also used for integration with an open capillary pump via microchannels to ensure a continuous flow over an assay chamber for glucose monitoring (Fig. 10b), however, this proof-of-concept only validated the fluidic connection and no on-chip analysis [158]. Furthermore, Jiang and Lillehoj connected a hollow microneedle array, consisting of SU-8 microneedles polymerized on a PET-substrate, with a test strip for the detection of a malaria biomarker, namely, *Plasmodium falciparum* histidine-rich protein 2 (PfHRP2) (Fig. 10c). By affixing a hydrophilic membrane to the hollow microneedle array, the ISF drawn into the micro-lumens was transferred onto the sample pad of the test strip. Lastly, the single existing proof-of-concept working with extracted blood was shown by Li and colleagues (Fig. 10d). They used a PDMS touch-switch, which had advanced from the previously mentioned PDMS vacuum chamber to a single hollow microneedle to pull 30 μL blood onto a polysulfone membrane for blood-serum separation. After filtration, the serum sample was drawn further through the

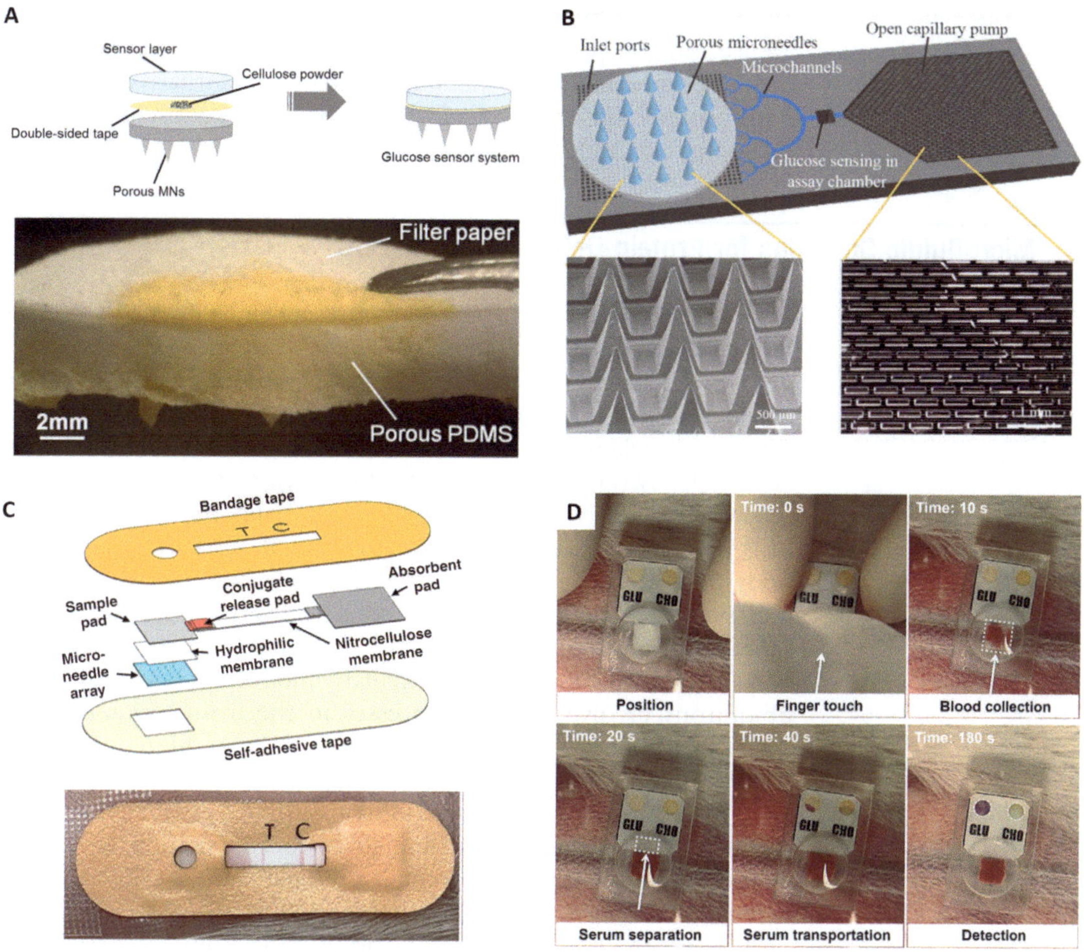

Fig. 10 Platforms integrating microneedle-based sample collection and paper-based testing by direct contact. (**a**) Porous PDMS microneedles affixed to a filter paper for single-step enzymatic colorimetric glucose testing. (**b**) Porous PDMS microneedles connected to an open capillary pump through microchannels to show the fluidic connection for continuous glucose monitoring. (**c**) Hollow microneedle array connected to a lateral flow test for PfHRP2 detection. The proof-of-concept was validated on piglet cadaver skin which was previously spiked with PfHRP2. (**d**) Single hollow microneedle for sampling blood actuated by a PDMS touch-switch and connected to a paper-based platform. The latter filtrated serum from whole blood in advance of loading detection pads for simultaneous glucose and cholesterol measurement. This device was tested on an in vivo rat model by sampling blood from the ear vein. Reproduced with permission from (**a**) Lee et al. ©2019 IEEE; (**b**) Takeuchi et al. ©2022 Springer; (**c**) Jiang et al. ©2020 Nature; (**d**) Li et al. ©2015 Royal Society of Chemistry

filter paper until it reached two detection pads for simultaneous glucose and cholesterol measurement [234].

Although all the above-mentioned examples illustrate the possibility to extract capillary blood or ISF directly from the body in a minimally invasive way, they (1) use a simple 1-step detection platform, (2) need manual addition of running buffer, or (3) lack control over sample volume. To increase the analytical and diagnostic reliability of these devices, they must be coupled to passive microfluidic systems with advanced fluid handling capability. Hence, sample processing can be performed immediately on the spot after its collection, enabling multistep quantitative assays for both protein and nucleic acid detection as will be discussed in the following two sections.

5 Microfluidic Solutions for Protein Detection at the POC

5.1 Proteins as Biomarkers

Molecular biomarkers are naturally present indicators of a specific biological state, typically associated with contraction of or stage of disease. Their detection and quantification allow the identification of physiological or pathological processes leading to a disease. These biomarkers exist in various forms, of which protein biomarkers are likely the most omnipresent in disease onset, response, and recovery [159]. Proteins function as essential parts of living organisms in multiple roles: providing structural support to cells and body tissues, transporting molecules, DNA replication, catalyzing metabolic reactions, and defending the body against foreign objects, among many others. The latter two, mainly comprising enzymes and antibodies, are particularly interesting because of their ubiquitous involvement in processes in the living organism and their morphological diversity, making them of high interest in POC diagnostics and the most used for protein analyses.

Protein detection in microfluidic systems generally involves specific recognition at a sensing zone located within a channel network. This recognition event is the final step in the sequential delivery of several reagents to this sensing zone, such as washing buffers, next to signal generation and amplification molecules (e.g., enzyme-linked immunosorbent assay (ELISA)). Additionally, to accommodating the assay process itself, integrated devices may also support sample collection and pretreatment steps beforehand, as well as a readout method (compatible with the application) afterward. Although some assays consist of fewer steps or can be completed in a single step, thus reducing complexity of the required platform (e.g., at-home pregnancy tests), many POC applications still rely on complex multistep assays. Efforts to develop both types of platforms for POC detection of protein biomarkers will be discussed below.

5.2 Porous Capillary Microfluidics for Protein Detection

Chromatographic LFAs are the simplest devices, consisting of a single strip of porous material in which the sample typically flows from one side to the other, i.e., from sample pad to waste pad. Most commercially available LFAs detect protein biomarkers with antibody bioreceptors, i.e., lateral flow immunoassays (LFIAs), as demonstrated by these commercially available examples: (1) a pregnancy test that detects human chorionic gonadotropin in urine, a hormone produced by the placenta during pregnancy [160]; (2) a

SARS-CoV-2 quick test that typically detects the viral nucleoprotein [161]; and (3) a rapid HIV test that detects the p24 antigen in blood or HIV antibody in saliva [162, 163]. In cases where a qualitative "yes or no" result suffices, and the assay can be optimized to 1 or 2 steps (i.e., sample and possibly buffer addition), LFIAs are perfect answers to POC diagnostic requirements. However, when multiple sequential steps need to be performed (e.g., sandwich ELISA), more flow control is required, which can be obtained in fluid networks in 2D or 3D μPADs. As touched upon before, research on μPADs is often still focusing on demonstrating technological abilities, such as flow paths in the channel network, assay optimization in the paper matrix, and novel detection methods, rather than developing completely integrated devices [164]. In this context, several advances are relevant to highlight, including the sample preparation, signal generation, or signal readout fronts.

First, several techniques for sample preparation have been demonstrated on μPADs [164]. As mentioned before, the most used biological sample is blood, in which a plethora of proteins can be found. Since red blood cells can interfere or inhibit analyte detection, the most common sample pretreatment in porous capillary microfluidics is removing red blood cells from blood to conduct the assay in plasma. Several methods have been demonstrated on μPADs, such as red blood cell agglutination [96] or aggregation [165], osmotic pressure [166], paper-based centrifugation [167], or most commonly, commercial filtration membranes integrated in the μPAD [168, 169]. Another relevant pretreatment step is metering the sample volume, as a controlled volume allows accurate analyte quantitation. Although metering is notoriously difficult in porous matrices, some efforts have shown robustness against sample volumes to a certain degree [170, 171].

Second, for analyte detection, several assay types have been reported on μPADs. The fluidic control for more complex assays is typically achieved through designing channel networks with specific features, which for example may include storing reagents in lyophilized state [100], programmable fluidic disconnects from a buffer tank [172], delay timers to increase incubation time [94, 173], and mechanical valves to enable sequential reagent delivery [173, 174]. Even more advanced fluidic control can be achieved when working in 3D, allowing execution of reagent steps upon folding [175–177] or sliding [177] parts of the device. The most common type of protein detection assays are immunoassays that combine the selective, antibody-based capture of the analyte with signal amplification, e.g., via enzymes (i.e., ELISA) [177, 178] or microparticles [179], although alternative bioreceptors such as molecularly imprinted polymers, aptamers, etc. have seen increasing interest. For example, aptamer-based recognition has been reported by Hui et al. [175], who proposed a μPAD with aptamer-based glutamate dehydrogenase detection to determine *Clostridium difficile* infection.

Finally, several methods for signal readout have been reported on μPADs [164], of which equipment-free readout is of most interest in POC settings. When a qualitative "yes or no" answer suffices, colorimetric readout via chromogenic enzyme reactions [117, 180], nanoparticle aggregation [179, 181] and pH indicators can be applied [117, 182]. Partial color signal quantification can be achieved by comparing the output color intensity to a standard, although the resolution is typically limited. Alternatively, further quantitation can be achieved by distance- or text-based readout. Distance-based readout relies on reading the length of a signal produced by interfacial interaction between paper and compounds [183], precipitation reactions [184, 185], or mobility influenced by reagent aggregation [184]. Despite their simplicity, these readout methods are susceptible to sample flow property variations and low precision. Alternatively, text-based readout provides results robust to user interpretation, although the proposed devices are still mostly qualitative (e.g., blood typing [186]) or semiquantitative [183]. When fully quantitative results are required, readout is often digital and done with portable devices. Small, dedicated devices such as glucose meters [183, 187] or smartphone-based readout [188, 189] are frequently used to automate signal readout further and decrease dependency on user interpretation. The introduction of screen-printed electrodes on μPADs has allowed electrochemical readout for protein biomarkers [176].

Overall, most of current μPADs still require multiple user intervention steps, such as manual sample preparation and multiple reagent additions, failing to achieve full integration. Nevertheless, some devices come very close to "sample in–result out" solutions including result quantification. Tian et al. [190] showed an integrated μPAD for cocaine detection in blood with aptamer-based target recognition, enzymatic signal amplification, and distance-based readout (Fig. 11a). User operation required sample addition and closing the device, yielding equipment-free readout in 30 min. Although the demonstrated application did not detect proteins, the device can be easily adapted to analytes for which a corresponding aptamer exists, or even to antibody-recognition. Meanwhile, Tenda et al. [171] reported a device for multiplex detection of anti-HIV1, anti-HA, and anti-DEN1 via bioluminescence resonance energy transfer switches and colorimetric signal generation (Fig. 11b). Here, user operation was limited to addition of a single blood droplet and taking a photograph with a camera 20 min later. Fu et al. [174] demonstrated a 2D μPAD with four shape-memory-polymer actuated fluid valves to quantify TNF-a in laryngeal tissue (Fig. 11c). The test was initiated by adding sample solution to the test zone and 250 μL buffer to the device inlet, upon which individual, dry-stored reagents were delivered to the

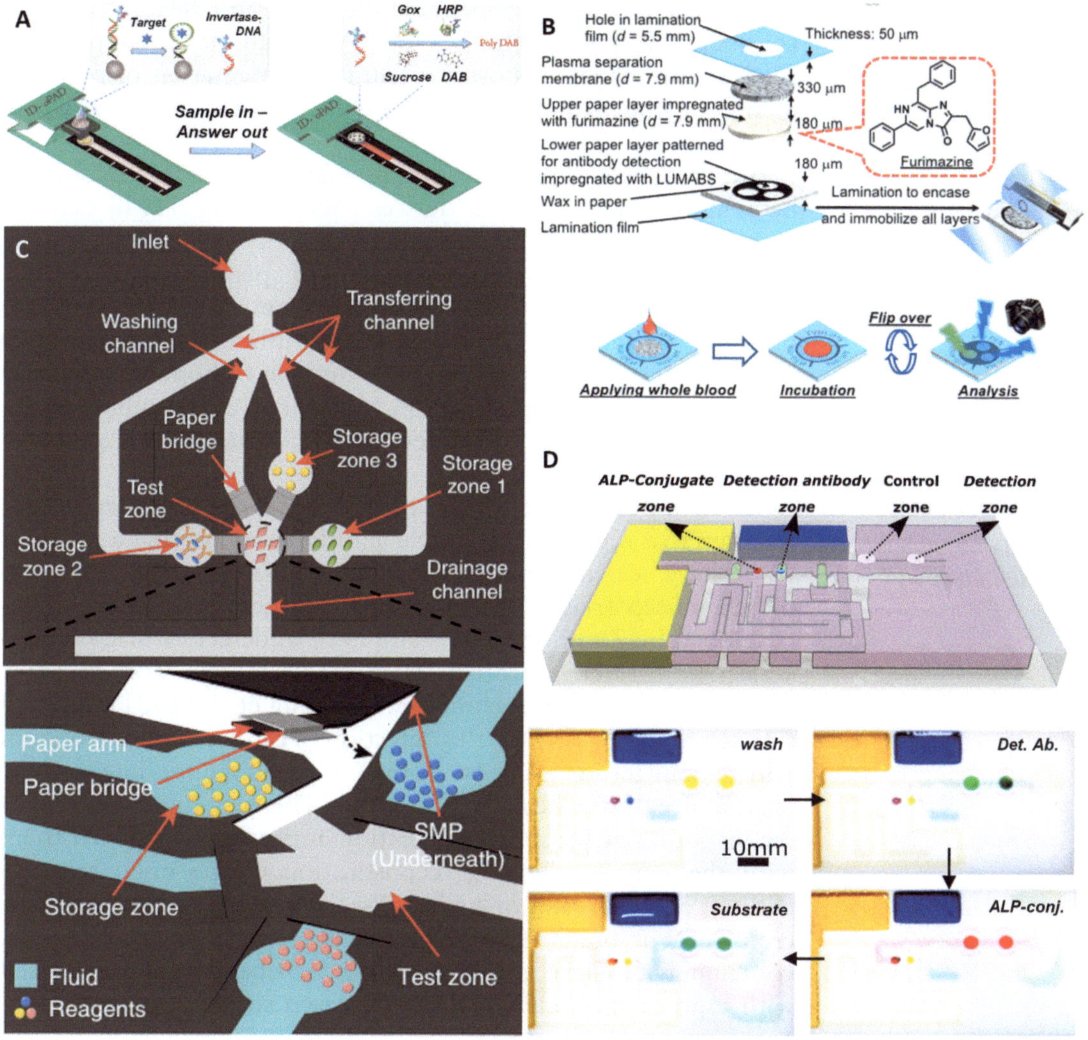

Fig. 11 Integrated porous capillary circuits. (**a**) Foldable μPAD with visual signal output based on aptamer/invertase-functionalized sepharose beads and distance-based readout. Abbreviations used: *GOx* glucose oxidase, *HRP* horseradish peroxidase, *DAB* 3,3′-diaminobenzidine. (**b**) Sample in–signal out μPAD with blood plasma separation, bioluminescence resonance energy transfer switches for target recognition, and colorimetric readout. (**c**) μPAD with shape-memory-polymer (SMP) actuated fluid valves for sequential reagent delivery. (**d**) 3D printed microfluidic device for ELISA, with dried reagent storage and two buffer tanks (yellow and blue), for washing/detection antibody/enzyme-conjugate and substrate delivery, respectively. Abbreviations used: *ALP* alkaline phosphatase, *Det. Ab.* detection antibody, *conj.* conjugate. Adapted with permission from (**a**) Tian et al. ©2017 ACS Publications; (**b**) Tenda et al. ©2018 Wiley; (**c**) Fu et al. ©2019 Wiley; and (**d**) Achille et al. ©2021 Wiley

test zone sequentially by operating the valves on a device which also conducted the colorimetric readout. Although these examples succeed in automatizing most steps, further integration of pretreatment/reagent loading steps and POC-compatible readout is still needed.

Interestingly, capillary-driven flow in a porous matrix can also be achieved in 3D-printed devices. Achille et al. [191] demonstrated a 3D printed, monolithic, capillarity-driven microfluidic device capable of timed sequence to perform IgE ELISA (Fig. 11d). The assay required user intervention via filling of two buffer tanks, and readout with a camera. Although the technology is promising to allow single-step production of homogeneous 3D geometries, further optimization is needed to streamline device fabrication and reduce the achieved 4 h 45 min time-to-result.

5.3 Microchannel Capillary Circuits for Protein Detection

In an attempt to improve their performance, simple 1-step immunoassays that only require sample addition, comparable to existing LFAs, have been implemented on solid capillary circuits. For example, the 1-step Triage® [55] chip with handheld readout device was developed by Zimmermann et al. [192] to detect cardiac biomarkers (i.e., C-reactive protein) from whole blood. More recently, further improving the technology, Hemmig et al. [193] presented a 1-step immunoassay for the detection of cardiac marker troponin I in human serum using two novel microfluidic modules: a self-coalescence and a bead lane module. The first one controllably reconstituted inkjet-spotted detection antibodies and mixed them with the sample, allowing reconstitution of the dried reagents without accumulation or typical dispersion profiles related to discrete reagent plugs. This then flowed to a bead lane module, containing capillary self-assembled microbeads (functionalized with capture antibodies), which increased surface area and could enable multiplex detection.

Although one-step assays can be improved with capillary circuits, even greater potential lies in developing devices capable of sequential reagent delivery in more complex assay protocols, enabling washing steps and signal amplification to increase the signal-to-noise ratio. Hereto, different capillaric elements can be combined, as has been demonstrated commonly in capillary circuits. For example, Ghosh et al. [188] presented a capillary platform using chemiluminescent detection of malaria biomarker PfHRP2 in plasma with a smartphone-based POCT. An injection-molded capillary circuit was designed and produced with a sample loading chamber to split the sample into two paths, test and control channels, and a capillary pump to power the circuit (Fig. 12a). The test and control channels contained lyophilized detection antibodies and chemiluminescence substrate, respectively, which were reconstituted by the sample and transported downstream. While

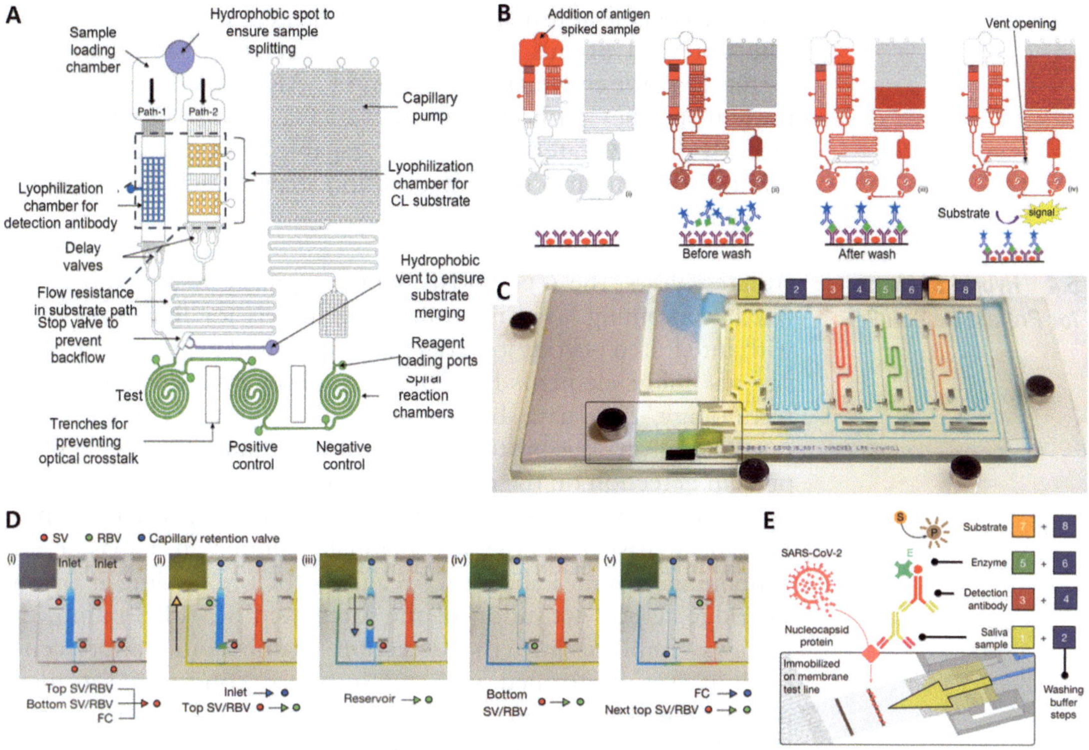

Fig. 12 Capillary microchannel circuits for multistep immunoassays. (**a**) Schematic of microchannel capillary flow assay for chemiluminescent detection of malaria biomarker and (**b**) step-by-step sequence showing liquid behavior (top) and assay working principle (bottom). (**c**) Assembled capillary circuit with integrated microfluidic chain reaction concept for automated detection of SARS-CoV-2 antibodies. (**d**) Operation of the different valving elements in the microfluidic chain reaction concept. Abbreviations used: *SV* stop valve, *RBV* retention burst valve, and *FC* functional connection. (**e**) Configuration of corresponding LFA. Adapted with permission, respectively, from (**a**, **b**) Ghosh et al. ©2020 Springer Nature; and (**c**–**e**) Yafia et al. ©2022 Springer Nature

the first path continued through the test channel and subsequent positive and negative control channels, the second path was delayed until it reached a stop valve (Fig. 12b). Upon opening of this valve by the user, the substrate also flowed through the test and control channels. Readout was performed in a smartphone-based handheld analyzer, using the smartphone as data processor, display device, and power source for the electronic circuit. Although the system demonstrated advanced integration, it still required manual opening of the hydrophobic air vent to initiate the flow of the chemiluminescence substrate, and plasma preparation was not yet integrated. Alternatively, Yafia et al. have developed a microfluidic chain reaction concept (Fig. 12c) [194]. The programmable action of capillary domino valves (consisting of a geometric stop, burst, and retention valves) was combined within a 3D-printed capillary circuit to enable automated sequential reagent delivery (Fig. 12d). The potential of this concept was illustrated by automating an 8-step LFA for enzymatic detection of SARS-CoV-2 nucleocapsid protein-specific antibodies in saliva (Fig. 12e). Overall, these recently demonstrated concepts are capable of advanced assay integration with minimal user intervention steps, which are promising for validation.

5.4 Hybrid Systems for Protein Detection

In latest years, the (i)SIMPLE hybrid capillary system was developed, combining solid microfluidic channels for fluid operations with integrated porous material pumps for fluid propagation. Since various fluidic elements have been established for this (i)SIMPLE technology, they have been combined to implement sequential reagent delivery for protein detection. As a first demonstration of protein detection on the platform, the Creasensor [114] was developed, a cartridge for creatinine detection in plasma samples. Herein, a plasma sample was manually diluted in a master mix containing enzymes and other reagents, before loading it on the cartridge. Upon activation of the integrated SIMPLE pump, the mix was imbibed into the detection chamber and fluorescence readout was performed with an in-house constructed readout device (Fig. 13a). More recently, two versions of an adalimumab (ADM) immunoassay have been implemented on the technology in a more integrated manner. The first cartridge performed single-step ADM detection with fiber-optic surface plasmon resonance (FO-SPR) readout (Fig. 13b) [195]. Herein, autonomous 1/20 mixing of plasma sample with reagents and delivery to a FO probe were demonstrated upon a single activation event. The FO-SPR readout was conducted on a benchtop readout device, yielding results in under 12 min. On the second cartridge, autonomous plasma separation from whole blood and a 5-step ADM ELISA in plasma with colorimetric readout was performed within 30 min (Fig. 13c) [196]. Overall, these concepts demonstrated the integration of a multitude of liquid operations and functionalities: (1) on-chip sample pretreatment, namely, plasma filtration from whole blood with elution into a channel by a SIMPLE pump, and 1/20 mixing of sample and reagent liquid plugs in the channel; (2) capture antibody immobilization on the cartridge material; (3) the autonomous execution of a multistep immunoassay, with washing and timed incubation steps; and (4) different readout strategies, including fluorescence and FO-SPR readout with in-house constructed, compact readout devices, as well as colorimetric readout with a handheld reflectometer. To yield completely integrated LOC devices, further efforts should focus on integration of sampling directly onto the cartridge (e.g., remote sampling with microneedles as discussed in Sect. 4), and fully integrated readout strategies, either with compact benchtop readout devices or highly portable solutions based on smartphones or naked-eye readout.

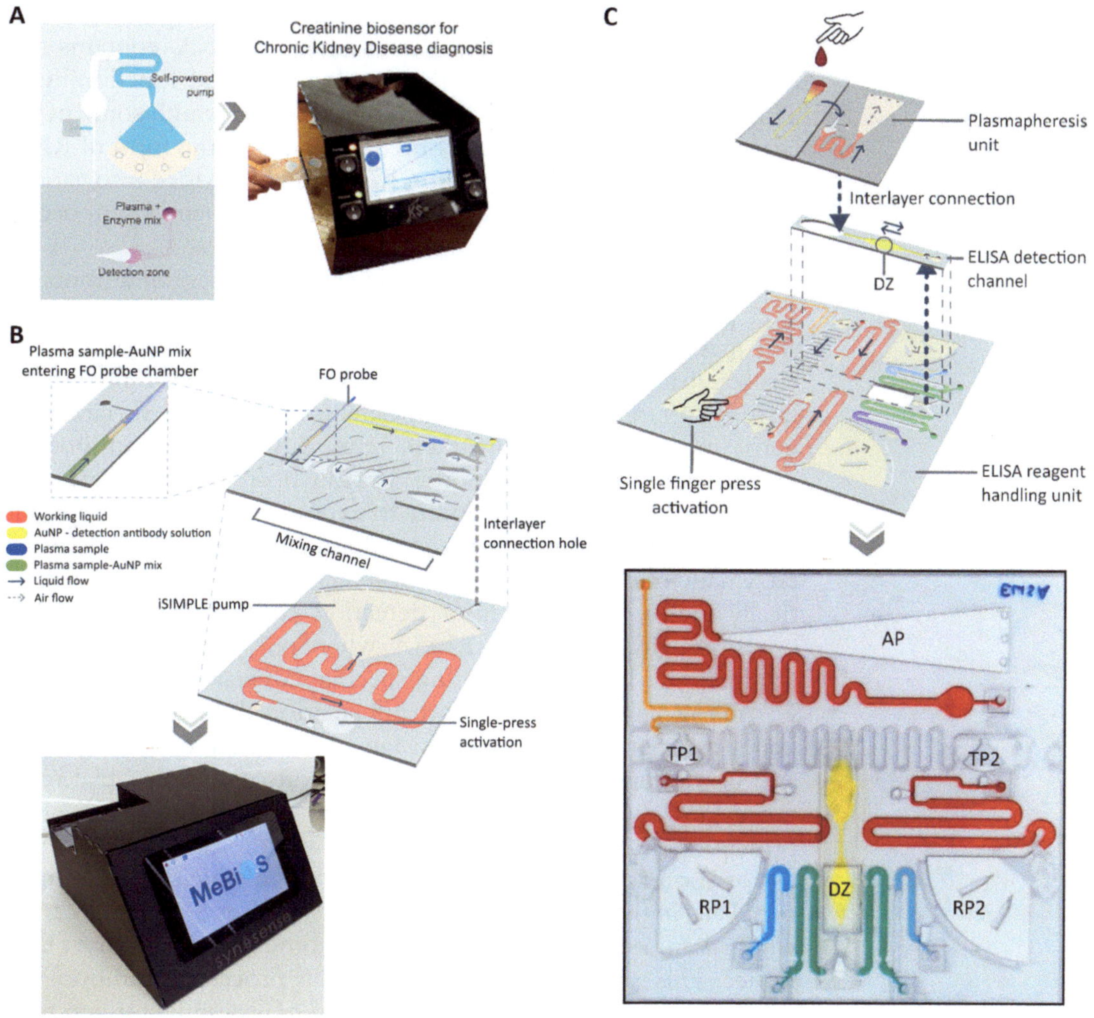

Fig. 13 Protein detection on (i)SIMPLE platform. (**a**) Creasensor for creatinine detection, consisting of a cartridge with SIMPLE pump and detection zone (left), readout with fluorescence on the self-constructed benchtop reader (right). (**b**) ADM-sensor, consisting of a cartridge with iSIMPLE pump capable of plasma sample (blue) and reagent (yellow) mixing and FO probe (top left). FO-SPR readout was conducted on the in-house constructed readout device. (**c**) Cartridge capable of plasmapheresis (top left) and five-step ADM ELISA from plasma sample (bottom left). Photograph of top view of chip filled with colored buffers for illustration purposes (right). Abbreviations used: *DZ* detection zone, *AP* activation pump, *TP* trigger pump, and *RP* reagent pump. Adapted with permission from (**a**) Dal Dosso et al. ©2018 Elsevier; (**b**) Qu & Ordutowski et al. ©2022 Elsevier; and (**c**) Ordutowski & Dal Dosso et al. ©2022 Elsevier

6 Microfluidic Solutions for Nucleic Acid Detection at the POC

6.1 Nucleic Acids as Biomarkers

Another important class of biomarkers are nucleic acids, molecules that carry the genetic information within the cell. Two main classes can be distinguished, comprising deoxyribonucleic acid (DNA) and ribonucleic acid (RNA), which only differ, structurally speaking, in the backbone sugar component (deoxyribose versus ribose) and a single nucleobase (thymine versus uracil). DNA contains the genetic information used in the development and functioning of all known living organisms and most viruses, although some of the latter also have their genetic content encoded in the form of RNA. Aside from carrying information, RNAs can be divided into various subtypes (mRNA, miRNA, etc.) depending on their origin, occurrence within the organism, and function.

In terms of biomarkers, nucleic acids offer several advantages over proteins. First, they allow for discriminating between active and past infections through direct pathogen detection (e.g., viruses, bacteria, fungi, and parasites) via its genetic fingerprint instead of targeting pathogen-specific proteins (e.g., surface antigens) or serological biomarkers (e.g., IgG or IgM antibodies) produced as immune response by the host (indirect pathogen detection). Moreover, the ability to exponentially multiply specific target sequences through the development of the PCR or other enzymatic nucleic acid amplification techniques (NAATs) has drastically increased the limit of detection compared to traditional immunological assays (i.e., ELISA) where only the generated signal is enhanced [197]. By targeting specific sequences in the genome, NAATs also provide much more information allowing to screen for genes associated with drug resistance, the latter becoming more and more important with the rising threat of antimicrobial resistance [199–201]. As NAATs are based on simple hybridization principles, assay development toward new pathogens goes relatively fast as soon as the target sequence is known, which is notoriously beneficial in the case of sudden outbreaks, such as SARS-COV2-19, where accurate diagnostic tests must become available in short time notice. This is in contrast with the lengthy process of immunological tests, which require the identification of antibody-antigen pairs with high affinity and specificity [201].

By enabling rapid and accurate nucleic acid quantification, PCR or other NAATs form the basis of modern well-established molecular diagnostic tools, which are implemented in most clinical laboratories [202]. Despite their proven analytical performance, their POC use has remained limited because they are skill-demanding and dependent on high-precision equipment for thermal cycling [198]. Moreover, to achieve the high sensitivity and selectivity for which NAATs are praised, they require multiple chemical or physical sample preparation steps to be performed. These involve: (1) lysis of the pathogen or other genetic carrier to extract the nucleic acid content; (2) purification of the nucleic acids by isolating them from inhibitory matrix compounds (e.g., hemoglobin), affecting downstream processes; (3) nucleic acid amplification, during which the target sequence is replicated with several orders of magnitudes; and (4) detection of the amplified nucleic acid, also referred to as amplicons. In the case of targeting free circulating nucleic acids (e.g., cancer detection), the first lysis step is not needed. Some promising examples of fully automated "near POC NAATs devices" such as the GeneXpert from Cepheid [203, 204] and Idylla from Biocartis [205] have already been introduced in the market. These systems typically use single-use disposable microfluidic cartridges in combination with a companion instrument that regulates the flow, temperature, and optical readout [207–209]. There is, however, a growing demand for affordable, fully autonomous chips that can support NAATs without depending on ancillary instrumentation.

Hereto, the emergence of isothermal NAATs might play a major role as these amplification assays operate at a constant temperature and therefore do not require sophisticated equipment for thermal cycling. This makes them much easier to integrate on standalone autonomous microfluidic chips. Furthermore, their analytical performance, in terms of sensitivity and specificity, was reported to be comparable with traditional PCR-based methods [209]. Some of the developed techniques also proved to (1) be more tolerant to inhibitory compounds, which can lead to decreased sensitivity or false-positive results, (2) have shorter time-to-results, and (3) be compatible with raw unprocessed samples without the need for purification [210, 211]. Over the past two decades, several isothermal NAATs (e.g., loop-mediated amplification (LAMP) and recombinase polymerase amplification (RPA)) have been developed, which are reviewed elsewhere [213–215]. Whereas the amplification step is simplified a lot by their constant temperature requirements, the development of self-driven sample in–result out NAATs still remains extremely challenging in terms of fluid operations. In particular, while immunological assays can be described as the sequential delivery of multiple sample and assay reagents over a fixed sensing region with immobilized capture molecules to form the eventual readout complex, NAATs involve the successive repetitive execution of multiple liquid handling steps (sample metering, reagent mixing, and incubation) on the sample liquid (see Fig. 14 for workflow comparison). In terms of microfluidics, the latter has proven to be of a much higher complexity and consequently, the number of fully autonomous and standalone sample-to-result NAATs has been limited to date.

6.2 Partially Integrated Microfluidic Solutions for Nucleic Acid Detection

Over the past years, researchers have sought to utilize simple, disposable, and cost-effective technical solutions to conduct each key step of the NAAT, while being compatible with low-resource POC settings. Paper and more in particular, μPADs, have been considered as an ideal platform to realize this by allowing complex multistep fluid handling. Moreover, the natural filtering action and chromatographic interactions of (nitro-)cellulose paper on the varying components within the sample were proven to be highly compatible with nucleic acid extraction. This was, for instance,

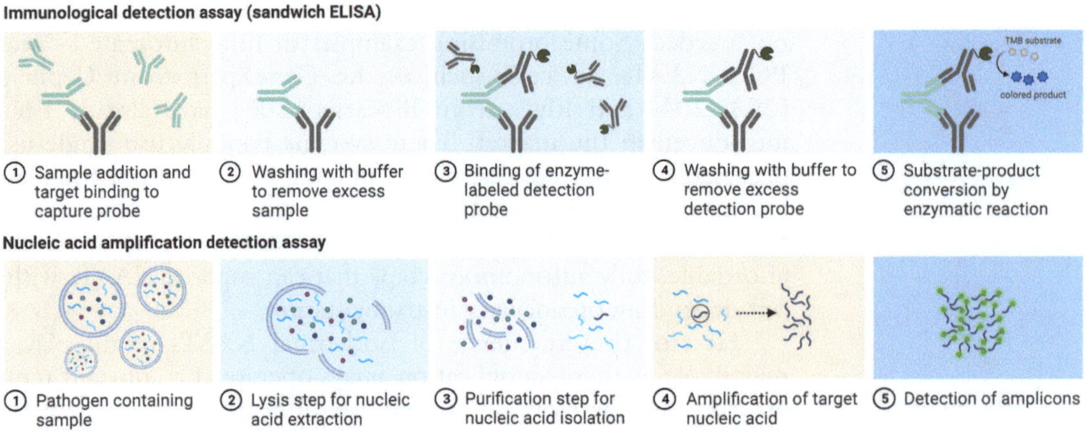

Fig. 14 Workflow comparison of the sequential sample preparation steps in immunological (sandwich ELISA) and nucleic acid amplification assays for the detection of proteins and nucleic acids, respectively. Created with BioRender.com

illustrated by Fronczek et al., who developed a 1D paper device to physically separate and purify the nucleic acid content of *Salmonella typhimurium* bacteria from clinical sample matrices [215]. To overcome the need for an additional lysis and dilution step prior to sample loading to the device, Govindarajan et al. demonstrated a 3D paper-based origami device with pre-stored dried reagents for *Escherichia coli* cell lysis and DNA extraction from highly viscous pig mucin samples through sequential folding of different paper fluidic circuits on top of each other [216]. Although the presented concepts illustrate the simplicity and efficacy of using paper for the extraction and purification of nucleic acids, the challenge remains on how to integrate these extraction modules with downstream amplification and detection methods toward sample in–result out POC systems.

Similar to protein detection, flow-through LFAs can also be used for the rapid detection of nucleic acids; however, they still require the sample preparation and amplification steps to be performed separately. Nevertheless, LFAs provide a low-cost, simple, and fast alternative to expensive laboratory-based readout methods. Two different types of nucleic acid-specific LFAs are typically used (i.e., NALFIA and NALFA, see further) depending on the biological recognition element for capturing and detecting the amplicon strands (Fig. 15a). In the NALFIA (nucleic acid lateral flow immunoassay), forward and reverse primers, tagged with antibody-specific proteins (e.g., biotin and FITC), are used within the amplification step to generate double-labeled amplicons. These form a sandwich assay by binding to their respective capture and detection antibodies, similar to traditional LFIAs for protein detection. Contrary to this, in the NALFA (nucleic acid lateral flow assay), amplicons with single-stranded tails are generated by

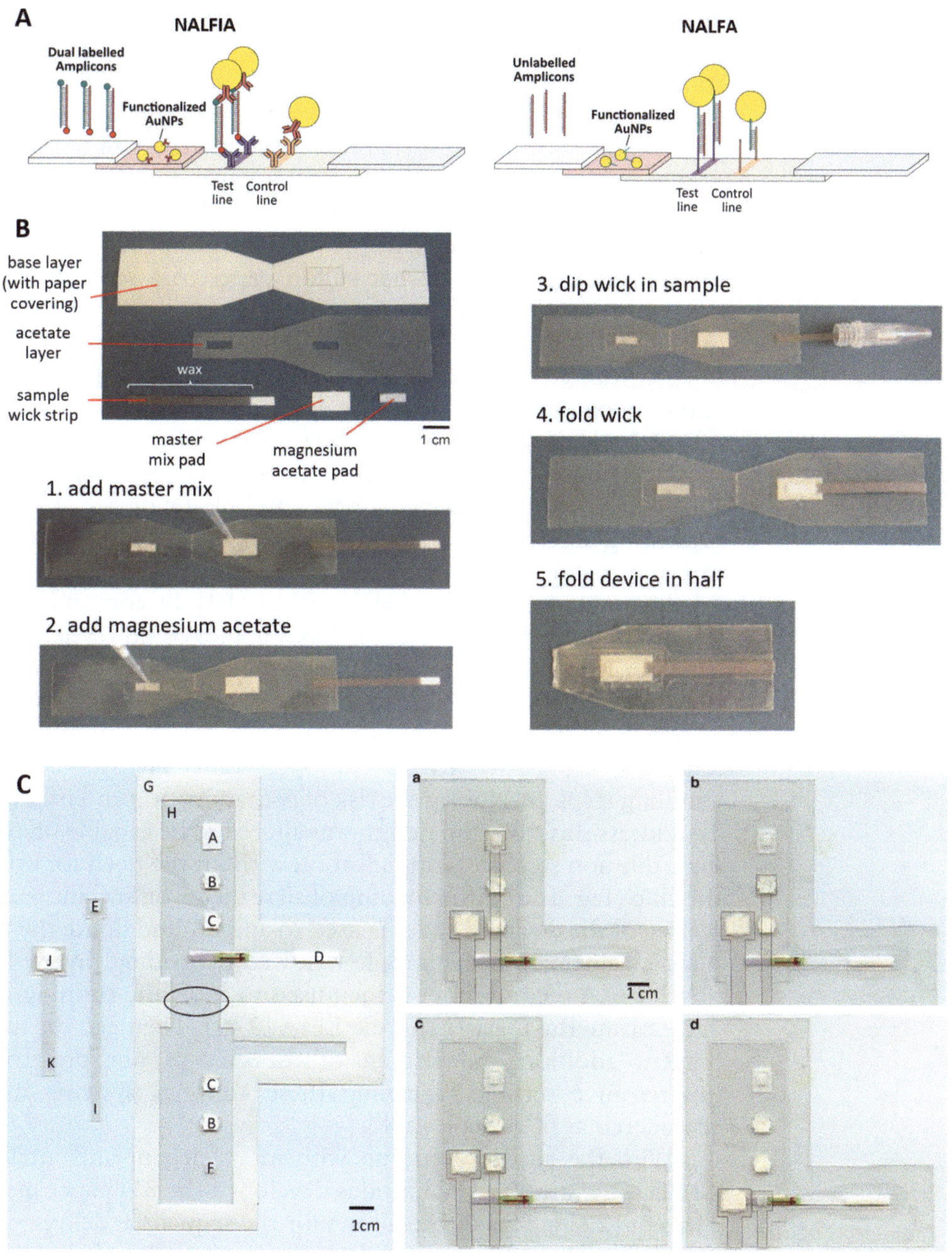

Fig. 15 Examples of partially integrated standalone NAATs. (**a**) Schematic of NALFIA and NALFA readout principles. (**b, c**) Overview of the different components and different operation steps of 2 paper fluidic sample in–result out devices integrating all steps of nucleic acid extraction/purification, amplification, and readout on LFIAs. Reproduced with permission from (**b**) Rohrman and Richards-Kortum. ©2012 Royal Society of Chemistry; (**c**) Cordray and Richards-Kortum. ©2015 Cordray and Richards-Kortum

incorporating extended primers. These hybridize to complementary single-stranded capture and reporter DNA probes to form a complex at the test line of the strip. Rodriguez and colleagues developed a paper-based assay that allowed the extraction and purification of Influenza A (H1N1) RNA from patient nasopharyngeal samples within a polyether sulfon paper matrix, followed by in-situ isothermal reverse transcriptase-LAMP target amplification and readout on separate NALFIA detection strips [217]. The need for transferring the polyether sulfon paper to a separated lateral flow strip was eliminated in a follow-up study, where they integrated all three modules (extraction, amplification, and detection) within a single paper-fluidic chip. By using this device, the authors demonstrated the direct detection of human papillomavirus 16 DNA in clinical patient cervical specimen in less than 1 h [218]. Inspired by the origami paper devices, Rohrman and Richards-Kortum developed a paper and plastic device (Fig. 15b), prestoring all essential reaction components (amplification master mix and magnesium acetate) to support an isothermal RPA assay for the on-chip detection of HIV DNA [219]. By correct folding of the different layers, as low as 10 DNA copies could be detected by using an LFA for readout. A huge drawback of this system is that the amplified product had to be diluted before application to the lateral flow strip in order to avoid false positive results, hence, requiring the device to be cut open. This involves additional handling steps, increasing the risk of contamination and user errors. To address this, the chip design was altered to be capable of executing a dilution step post-amplification without the need for opening the chip (Fig. 15c). After loading of all reagents on specific pads and sealing of the device, a slider is used to amplify and dilute the target nucleic acids within the sample followed by loading on the lateral flow strip. A second slider is then used to apply the running buffer and activate the malaria-specific LFIA [220]. The need for manual reagent addition and sealing of the chip is not preferred in low-resource settings, meaning these kinds of systems are not considered as field-deployable.

With the aim to come up with more elegant and integrated NAATs, the group of Whitesides developed the 3D paper machine concept [221]. This device allows for the sequential delivery of the sample, wash buffers, amplification master mix, and detection reagents with minimal pipetting and user-intervention requirements by sliding a paper strip in and out of a fluidic path, as illustrated in Fig. 16a. Consequently, all steps of an isothermal LAMP assay detecting the *Escherichia coli* malB gene were integrated on this handheld single-use disposable POC device, showing a limit of detection of five cells. A drawback of the system is that it depends on a UV light for readout and target quantification. Another attempt toward complete system integration was

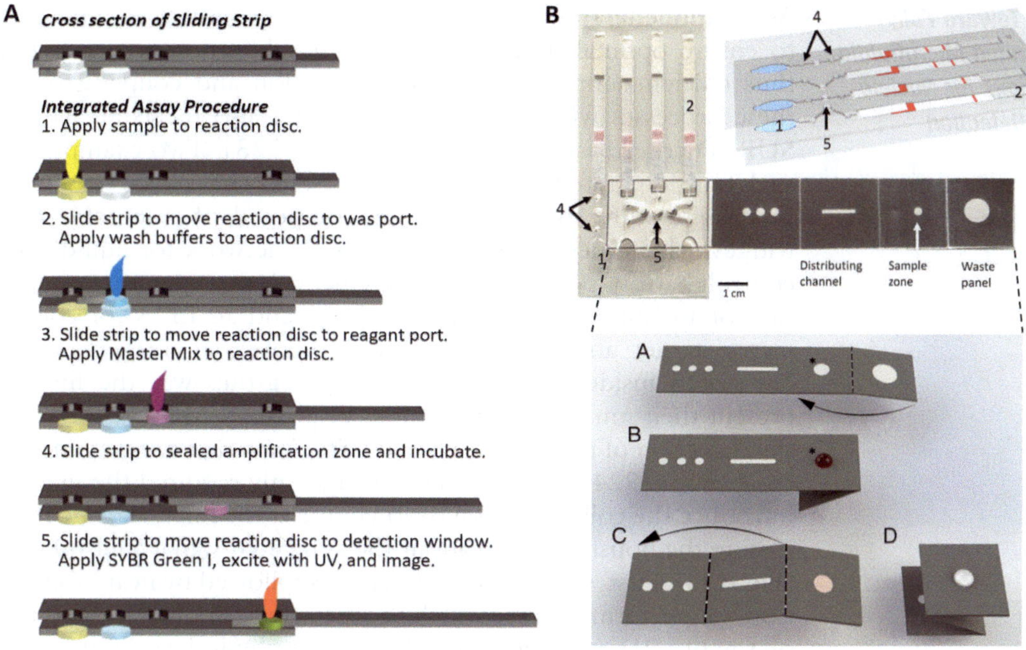

Fig. 16 (**a**) 3D paper machine concept ensuring sequential reagent delivery of a LAMP assay through a manual sliding element. (**b**) Combination of 3D origami foldable paper device enabling the sample pre-processing of a LAMP assay for the multiplex detection of malaria subspecies on LFIAs. Reproduced with permission from (**a**) Connelly et al. ©2015 American Chemical Society; B) Reboud et al. ©2019 Reboud et al. Published by PNAS

realized by Reboud et al. They coupled their previously established wax-patterned paper origami device, incorporating all steps of DNA extraction and processing to accommodate an isothermal LAMP, with a plastic lateral flow diagnostic device holding four strips for the multiplex detection of malaria in blood (Fig. 16b) [92, 222]. Field validation studies, performed in village schools of Uganda, illustrated the capability of the device for the fast (<50 min) diagnosis of the different malaria species in patients. This study also showed the high sensitivity and specificity of the test as 98% of the malaria-infected patients could be detected.

While the NAATs presented here integrate all steps for sample preparation, amplification, and detection, multiple manual handling steps (i.e., following a specific sequence for reagent addition and paper folding) still must be performed. This limits their implementation in the field and usage by untrained end-users. Moreover, for generating the heat, essential for the amplification assay, active heating elements such as a heat block, gas stove, or hot plate were used. In settings without ensured access to electricity, this is not ideal and alternative solutions such as battery-powered systems or exothermic chemical reactions are favored.

6.3 Toward Fully Integrated Microfluidic Solutions for Nucleic Acid Detection

As mentioned in the previous section, a heat source is still essential to ensure optimal functioning of the amplification enzymes and thus efficient amplification. The integration and coupling of all these processes within a single device that also meets all REASSURED criteria is very challenging [209, 223]. Two examples of such fully integrated sample-to-answer NAATs have been presented in the literature. Both devices are composed of disposable cartridges, which incorporate a paper fluidic network for liquid flow control, battery-powered heating elements for temperature regulation of an isothermal amplification step, and an LFIA for colorimetric target amplicon detection. Moreover, all reagents were dry-stored inside the device. The Yager group was the first to present their multiplex autonomous disposable (MAD) NAAT for the detection of methicillin-resistant *Staphylococcus aureus* genes in nasal swabs [224]. This simple cartridge only required the user to introduce the collected nasal swab sample at the inlet, activate it by closing the housing, and readout the generated signal after 60 min (Fig. 17a). First, enzymatic bacterial lysis followed by heat inactivation was performed, after which the extracted sample was delivered on top of a 2D paper network where it was physically split over two dried reagent pads as depicted in the close-up of Fig. 17a. Here, the lysate was reconstituted with internal controls and dried master mixtures of an isothermal strand displacement amplification assay (i.e., primers, enzymes, etc.). The physical separation of the lysate allowed it to target two different genes and discriminate between methicillin-susceptible (ldh1-gene) and methicillin-resistant (mecA-gene) species. After a 30 min heating step at 50 °C, the amplified product was automatically loaded on the two LFIAs, containing four capture lines: a target line for each gene and two internal positive controls indicating successful release of the gold nanoparticles and successful performance of amplification step. To control the sample delivery from the sample chamber to the 2D paper network and the amplification to the detection zone, wax-based valves were used, which opened upon applying heat. A very similar concept was developed by Tang et al., who demonstrated a 3D-printed disposable device integrating three separate modules for the extraction, amplification, and detection of nucleic acids (Fig. 17b) [225]. Sponge-based reagent reservoirs connected to a paper-based fluidic network were utilized for nucleic acid extraction, while a battery-powered ultrathin heater was used to support a helicase-dependent amplification assay. By means of a manual slider, the amplified product was eventually delivered to a lateral flow strip for nucleic acid detection. The device was validated to detect *Salmonella typhimurium* within spiked wastewater, milk, juice, and eggs with good sensitivity (limit of detection of 10^2, 10^3, 10^3, and 10^2 CFU/mL^{-1}) while proven to have high specificity against other species (*Vibrio parahaemolyticus*, *Listeria monocytogenes*, *Escherichia coli*, and *Staphylococcus aureus*).

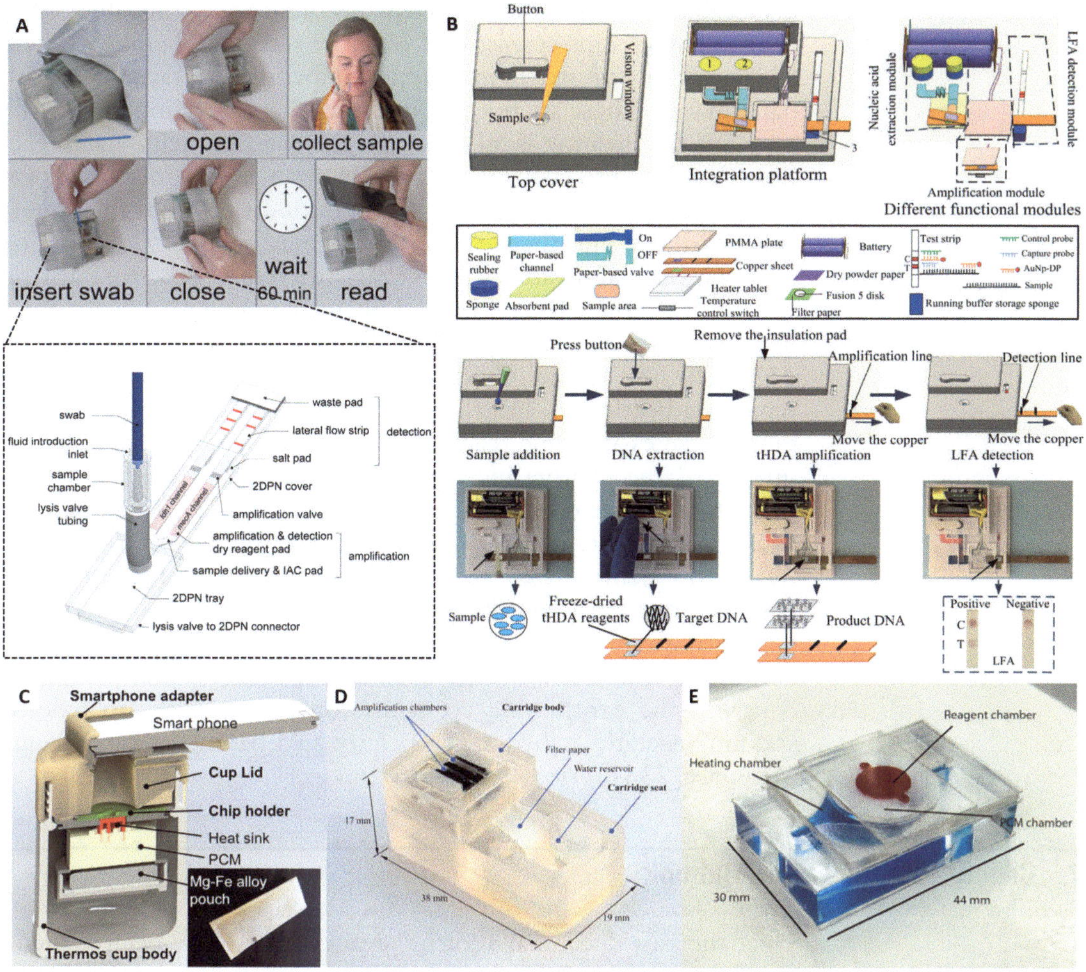

Fig. 17 Fully integrated NAATs. (**a**) Stepwise workflow for operating the MAD NAAT and a close-up of the internal sample inlet and 2D paper network with all indicated components. (**b**) Schematic of the integrated NAAT cartridge with its different functional modules (top) and the working principle of how the different steps of the NAAT are performed (bottom). (**c**) Example of the thermal cup device for passive heat generation and (**d**, **e**) microfluidic systems with an integrated passive heat source. Reproduced with permission from (**a**) Lafleur et al. ©2016 The Royal Society of Chemistry; (**b**) Tang et al. ©2017 The Royal Society of Chemistry; (**c**) Liao et al. ©2016 Elsevier; (**d**) Liu et al. ©2011 The Royal Society of Chemistry; (**e**) Vloemans et al. ©2020 Elsevier

Despite their multiple advantages and full integration, both described concepts rely on battery-powered resistors for heat generation and thermal control, which is not preferred in terms of electronic complexity, cost, and waste management. To eliminate the need for electronic circuitry and batteries, electricity-free heat sources such as toe warmers or chemical heating principles have been exploited to drive the amplification reaction [226, 227]. For instance, Liao et al. introduced the thermal smart-cup concept with controlled chemical heating (i.e., magnesium–iron exothermic corrosion reaction) to drive the isothermal amplification assay in a disposable microfluidic cassette, as illustrated in Fig. 17c [228]. These chips consist of 1 or more reaction chambers where all reagents are prestored in lyophilized form and encapsulated with paraffin wax. These are released into the reaction chamber from the moment that heating is initiated and the paraffin melts. Each reaction chamber also incorporates a Whatman FTA® membrane for nucleic acid extraction, isolation, and purification from the sample (Fig. 17d, e). The thermal cup format has become very popular over the years and various types with different exothermic driving sources, with or without coupled phase change materials for thermal regulation, have been demonstrated to perform well [230–232]. The lack of microfluidic integration and scalability, however, remains a bottleneck and therefore researchers attempted to integrate these exothermic reactions directly on chip. Liu and colleagues used the exothermic reaction between an Mg-Fe alloy and water to generate on-chip heating (Fig. 17d) [232]. They used the wicking properties of filter paper to control the flow rate at which water was delivered to the Mg-Fe alloy and consequently the reaction rate and amount of generated heat. An alternative approach was presented by Vloemans et al. (Fig. 17e), who realized controlled localized on-chip heating between 30 and 50 °C by coupling the exothermic crystallization reaction of supercooled sodium acetate trihydrate to fatty-acid-based phase change materials [233].

7 Conclusion and Future Outlook

Over the past decennia, a lot of advancements have been made in the field of self-powered microfluidics with the aim to integrate laboratory-based detection assays on fully autonomous and standalone platforms. By doing so, they promise to deliver simple yet reliable diagnostic tools for resource-limited settings, outside centralized clinical facilities, hereby increasing the widespread accessibility to proper healthcare in both developed and developing countries. The latter is essential to address the expected challenges the global healthcare systems will be facing as a result of the growing and aging world population, increased globalization, and climate change, along with continuously reducing national healthcare budgets.

The field of self-powered microfluidics itself is mainly dominated by paper- and microchannel-based microfluidic principles, illustrating the potential of capillary action as a passive driving source for liquid manipulation. The continuously improving fabrication processes enable more precise control over the geometric and chemical properties of either the continuous microchannels or the porous matrices. Hence, specific flow events can be programmed within the microfluidic circuit of these capillary-driven systems. This has resulted in the development of advanced capillary platforms capable of performing complex flow operations such as sequential reagent delivery to mitigate multistep immunological detection assays. However, realizing self-powered paper- or channel-based microfluidic solutions for supporting sample in–result out NAATs has proven to be much more challenging as these require more complicated and successive sample preparation steps to be performed. Despite the advantage of using capillary action as "direct" driving source of liquid manipulation, this also introduces two fundamental challenges limiting its overall potential:

1. The dependency of the flow behavior on the interaction between the "to be manipulated liquid" (i.e., blood sample) and "transport medium" (i.e., channel walls, pours substrate) makes the system very sensitive to variations in sample matrix properties such as viscosity and surface tension. This variability is typically observed in biological samples like whole blood (e.g., person-to-person variability of hematocrit level) and might result in a completely different flow behavior and even failure of certain flow elements (e.g., valves).
2. The flow in capillary systems is determined by the capillary pressure drop over the advancing and receding liquid–air interface. Consequently, it is very difficult to manipulate confined liquid plugs through the whole microfluidic network. Likewise, serial fluid operations of liquid splitting and merging are extremely challenging to integrate on these systems. This limits its potential for bioassay applications that require multiple consecutive sample pre-processing steps (e.g., lysis, DNA purification, amplification, etc.).

In this context, the authors believe that hybrid paper- and channel-based solutions hold a lot of potential for supporting NAATs as in these systems the liquid manipulation can be driven by "indirect" driving sources within the microfluidic circuit. Another advantage of this hybrid technology is the ability to program precise flow control within the microchannel circuit without the need for complex treatment steps. Nevertheless, further improvements to achieve better control over the porous substrate parameters (i.e., pore size, porosity) are required to reduce flow rate variability. This remains an inherent problem with paper-based microfluidics to date. In this context, advancements in fabrication methods such as 3D printing might offer a solution in the near future.

Despite the numerous self-powered LOC devices proven to be successful in controlled laboratory settings, the number of devices that reach the market remains limited. This might be the result of a continuous search for new or improving existing technical microfluidic solutions to increase, among others, specificity and sensitivity. However, a more holistic approach should be taken, starting from the early phases of the development process of microfluidic devices to include more of the REASSURED criteria. Indeed, challenges that are nowadays often overlooked are mostly associated with user-experience. Self-powered, fully integrated POCTs including minimally invasive sample collection upon a single activation step are still lacking. In particular, microfluidic engineers should address more practical problems such as the storage and release of multiple liquid reagents without additional manual interventions being required. Another overlooked aspect is the user-interface (e.g., deposition of the sample) to make the use of the device intuitive.

Nevertheless, the authors believe that over the past few years, the vision of the POC microfluidic device development approach has shifted from a purely technological-driven perspective toward a more end-user-oriented approach. In this aspect, the COVID-19 pandemic can be seen as a real game-changer. While it provided many medical device developers with the opportunity to test their novel POC systems in the field already in an early phase, it also confronted them with the many practical issues encountered from the end-user side. As a consequence, it forced them to go back to the drawing table to redesign and simplify their systems. To summarize, the field of self-powered microfluidics has advanced a lot over the past few years, on the one hand, by the development of innovative technological aspects that enable more complex fluid manipulations, whereas on the other hand, the development approach has become more patient-centric. This combination is very promising as it is expected to lead toward more commercially viable integrated POCT, which effectively makes it to the market.

References

1. Atella V et al (2019) Trends in age-related disease burden and healthcare utilization. Aging Cell 18:1–8
2. Baker RE et al (2021) Infectious disease in an era of global change. Nat Rev Microbiol. https://doi.org/10.1038/s41579-021-00639-z
3. Billings PR (2006) Three barriers to innovative diagnostics. Nat Biotechnol 24:917–918
4. Rohr UP et al (2016) The value of in vitro diagnostic testing in medical practice: a status report. PLoS One 11:1–16
5. Kumar S et al (2021) Aspects of point-of-care diagnostics for personalized health wellness. Int J Nanomedicine 16:383–402
6. Peeling RW, Mabey D (2010) Point-of-care tests for diagnosing infections in the developing world. Clin Microbiol Infect 16:1062–1069
7. Yager P, Domingo GJ, Gerdes J (2008) Point-of-care diagnostics for global health. Annu Rev Biomed Eng 10:107–144
8. Mabey D, Peeling RW, Ustianowski A, Perkins MD (2004) Diagnostics for the developing world. Nat Rev Microbiol 2:231–240
9. Bottero J et al (2017) Simultaneous human immunodeficiency virus- hepatitis B-hepatitis C point-of-care tests improve outcomes in linkage-to-care: results of a randomized control trial in persons without healthcare coverage. Open Forum Infect Dis 2:2633851
10. Govender K, Parboosing R, Siyaca N, Moodley P (2016) Dried blood spot specimen quality and validation of a new pre-analytical processing method for qualitative HIV-1 PCR, KwaZulu-Natal, South Africa. Afr J Lab Med 5:1–7
11. Ayukekbong JA, Ntemgwa M, Atabe AN (2017) The threat of antimicrobial resistance

in developing countries: causes and control strategies. Antimicrob Resist Infect Control 6:1–8

12. Byarugaba DK (2004) Antimicrobial resistance in developing countries and responsible risk factors. Int J Antimicrob Agents 24:105–110
13. Udugama B et al (2020) Diagnosing COVID-19: the disease and tools for detection. ACS Nano 14:3822–3835
14. Desai AN et al (2019) Real-time epidemic forecasting: challenges and opportunities. Health Secur 17:268–275
15. Hansen GT (2020) Point-of-care testing in microbiology: a mechanism for improving patient outcomes. Clin Chem 66:124–137
16. Kelly-Cirino CD et al (2019) Importance of diagnostics in epidemic and pandemic preparedness. BMJ Glob Health 4:1–8
17. St John A, Price CP (2014) Existing and emerging Technologies for Point-of-Care Testing. Clin Biochem Rev 35:155–167
18. Larsson A, Greig-Pylypczuk R, Huisman A (2015) The state of point-of-care testing: a european perspective. Ups J Med Sci 120:1–10
19. Lei BUW, Prow TW (2019) A review of microsampling techniques and their social impact. Biomed Microdevices 21:1–30
20. Lim MD (2018) Review article: dried blood spots for global health diagnostics and surveillance: opportunities and challenges. Am J Trop Med Hyg 99:256–265
21. Zakaria R, Allen KJ, Koplin JJ, Roche P, Greaves RF (2016) Advantages and challenges of dried blood spot analysis by mass spectrometry across the total testing process. Ejifcc 27:288–317
22. Tanna S, Lawson G (2015) Self-sampling and quantitative analysis of DBS: can it shift the balance in over-burdened healthcare systems? Bioanalysis 7:1963–1966
23. Sikombe Id Ket al (2019) Accurate dried blood spots collection in the community using non-medically trained personnel could support scaling up routine viral load testing in resource limited settings. https://doi.org/10.1371/journal.pone.0223573
24. Doornekamp L et al (2020) Dried blood spot cards: a reliable sampling method to detect human antibodies against rabies virus. PLoS Negl Trop Dis 14:1–10
25. Toh ZQ et al (2021) The use of dried blood spots for the serological evaluation of SARS-CoV-2 antibodies. J Public Health (Oxf):1–4. https://doi.org/10.1093/pubmed/fdab011
26. Wu G, Zaman MH (2012) Low-cost tools for diagnosing and monitoring HIV infection in low-resource settings. Bull World Health Organ 90:914–920
27. Xing J, Loureiro J, Patel MT, Mikhailov D, Gusev AI (2020) Evaluation of a novel blood microsampling device for clinical trial sample collection and protein biomarker analysis. Bioanalysis 12:919–935
28. Törnqvist E et al (2014) Strategic focus on 3R principles reveals major reductions in the use of animals in pharmaceutical toxicity testing. PLoS One 9:1–11
29. Urdea AM et al (2006) Requirements for high impact diagnostics. Nature 444:73–79
30. Hay Burgess DC, Wasserman J, Dahl CA (2006) Global health diagnostics. Nature 444(Suppl):1–2
31. Schito M et al (2012) Opportunities and challenges for cost-efficient implementation of new point-of-care diagnostics for HIV and tuberculosis. J Infect Dis 205:169–180
32. Clark LC, Lyons C (1962) Electrode systems for continuous monitoring in cardiovascular surgery. Ann N Y Acad Sci 102:29–45
33. Ehrenkranz JRL (2002) Home and point-of-care pregnancy tests: a review of the technology. Epidemiology 13:15–18
34. Peeling RW, Holmes KK, Mabey D, Ronald A (2006) Rapid tests for sexually transmitted infections (STIs): the way forward. Sex Transm Infect 82:1–6
35. Land KJ, Boeras DI, Chen XS, Ramsay AR, Peeling RW (2019) REASSURED diagnostics to inform disease control strategies, strengthen health systems and improve patient outcomes. Nat Microbiol 4:46–54
36. Chen XS (2013) Rapid diagnostic tests for neurosyphilis. Lancet Infect Dis 13:918–919
37. Jain S et al (2021) Internet of medical things (IoMT)-integrated biosensors for point-of-care testing of infectious diseases. Biosens Bioelectron 179:113074
38. Paul S et al (2021) Industry 4.0 applications for medical/healthcare services, pp 1–32
39. Smith S, Korvink JG, Mager D, Land K (2018) The potential of paper-based diagnostics to meet the ASSURED criteria. RSC Adv 8:34012–34034
40. Manz A, Graber N, Widmer HM (1990) Miniaturized total chemical analysis systems: a novel concept for chemical sensing. Sensors Actuators 1:244–248
41. Mark D, Haeberle S, Roth G, von Stetten F, Zengerle R (2010) Microfluidic lab-on-a-chip

platforms: requirements, characteristics and applications. Chem Soc Rev 39:1153

42. Beebe DJ, Mensing GA, Walker GM (2002) Physics and applications of microfluidics in biology. Annu Rev Biomed Eng 4:261–286
43. van Lintel HTG, van De Pol FCM, Bouwstra S (1988) A piezoelectric micropump based on micromachining of silicon. Sensors Actuators 15:153–167
44. Verpoorte EMJ et al (1994) Three-dimensional micro flow manifolds for miniaturized chemical analysis systems. J Micromech Microeng 4:246–256
45. Duffy DC, McDonald JC, Schueller OJA, Whitesides GM (1998) Rapid prototyping of microfluidic systems in poly(dimethylsiloxane). Anal Chem 70:4974–4984
46. Xia Y, Whitesides GM (1998) Soft lithography. Annu Rev Mater Sci 28:153–184
47. Temiz Y, Lovchik RD, Kaigala GV, Delamarche E (2015) Lab-on-a-chip devices: how to close and plug the lab? Microelectron Eng 132:156–175
48. Volpatti LR, Yetisen AK (2012) Commercialization of point-of -care diagnostic microfluidic devices. Lab Chip 12:2118–2134
49. Mohammed MI, Haswell S, Gibson I (2015) Lab-on-a-chip or chip-in-a-lab: challenges of commercialization lost in translation. Procedia Technol 20:54–59
50. Xu L, Wang A, Li X, Oh KW (2020) Passive micropumping in microfluidics for point-of-care testing. Biomicrofluidics 14:031503
51. Narayanamurthy V et al (2020) Advances in passively driven microfluidics and lab-on-chip devices: a comprehensive literature review and patent analysis. RSC Adv 10:11652–11680
52. Young T (1805) An essay on the cohesion of fluids. Philos Trans 95:65–87
53. Laplace P-S (1805) Traité de Mécanique Céleste. Supplements au Livre X
54. Chen T, Chiu MS, Weng CN (2006) Derivation of the generalized young-laplace equation of curved interfaces in nanoscaled solids. J Appl Phys 100:1–5
55. Olanrewaju A, Beaugrand M, Yafia M, Juncker D (2018) Capillary microfluidics in microchannels: from microfluidic networks to capillaric circuits. Lab Chip 18:2323–2347
56. Bruus H (2006) Lecture notes theoretical microfluidics. Physics
57. Delamarche E, Bernard A, Schmid H, Michel B, Biebuyck H (1997) Patterned delivery of immunoglobulins to surfaces using microfluidic networks. Science 80: 276, 779–781
58. Safavieh R, Juncker D (2013) Capillarics: pre-programmed, self-powered microfluidic circuits built from capillary elements. Lab a chip – miniaturisation. Chem Biol 13:4180–4189
59. Park J, Han DH, Park JK (2020) Towards practical sample preparation in point-of-care testing: user-friendly microfluidic devices. Lab Chip 20:1191–1203
60. Safavieh R, Tamayol A, Juncker D (2015) Serpentine and leading-edge capillary pumps for microfluidic capillary systems. Microfluid Nanofluidics 18:357–366
61. Temiz Y, Delamarche E (2018) Sub-nanoliter, real-time flow monitoring in microfluidic chips using a portable device and smartphone. Sci Rep 8:1–12
62. Gervais L, Delamarche E (2009) Toward one-step point-of-care immunodiagnostics using capillary-driven microfluidics and PDMS substrates. Lab Chip 9:3330–3337
63. Zimmermann M, Schmid H, Hunziker P, Delamarche E (2007) Capillary pumps for autonomous capillary systems. Lab Chip 7: 119–125
64. Olanrewaju AO, Robillard A, Dagher M, Juncker D (2016) Autonomous microfluidic capillaric circuits replicated from 3D-printed molds. Lab Chip 16:3804–3814
65. Zimmermann M, Hunziker P, Delamarche E (2008) Valves for autonomous capillary systems. Microfluid Nanofluidics 5:395–402
66. Andersson H, Van Wijngaart W, Der G, P. & Niklaus, F. (2001) Hydrophobic valves of plasma deposited octafluorocyclobutane in DRIE channels. Sensors Actuators B Chem 75:136–141
67. Kazemzadeh A, Ganesan P, Ibrahim F, He S, Madou MJ (2013) The effect of contact angles and capillary dimensions on the burst frequency of super hydrophilic and hydrophilic centrifugal microfluidic platforms, a CFD study. PLoS One 8:1–12
68. Convery N, Gadegaard N (2019) 30 years of microfluidics. Micro Nano Eng 2:76–91
69. Martinez AW et al (2010) Programmable diagnostic devices made from paper and tape. Lab Chip 10:2499–2504
70. Martinez AW, Phillips ST, Whitesides GM (2008) Three-dimensional microfluidic devices fabricated in layered paper and tape. Proc Natl Acad Sci USA 105:19606–19611
71. Martinez AW, Phillips ST, Butte MJ, Whitesides GM (2007) Patterned paper as a platform for inexpensive, low-volume, portable bioassays. Angew Chem Int Ed 46:1318–1320

72. Nilghaz A, Ballerini DR, Shen W (2013) Exploration of microfluidic devices based on multi-filament threads and textiles: a review. Biomicrofluidics 7:1–16
73. Xing S, Jiang J, Pan T (2013) Interfacial microfluidic transport on micropatterned superhydrophobic textile. Lab Chip 13: 1937–1947
74. Safavieh R, Zhou GZ, Juncker D (2011) Microfluidics made of yarns and knots: from fundamental properties to simple networks and operations. Lab Chip 11:2618–2624
75. Gong MM, Sinton D (2017) Turning the page: advancing paper-based microfluidics for broad diagnostic application. Chem Rev 117:8447–8480
76. Washburn EW (1921) The dynamics of capillary flow. Phys Rev 18:206–209
77. Lucas R (1918) Rate of capillary Ascension of liquids. Kolloid 23
78. Darcy (1856) Les Fontaines Publiques de La Ville de Dijon: Exposition et Application des Principes a Suivre et des Formulesa Employer dans les Questions de Distribution d'Eau
79. Datta AK (2002) Chapter 6: convection heat transfer. In: Biological and bioenvironmental heat and mass transfer. CRC Press, Boca Raton, pp 95–121
80. Camplisson CK, Schilling KM, Pedrotti WL, Stone HA, Martinez AW (2015) Two-ply channels for faster wicking in paper-based microfluidic devices. Lab Chip 15:4461–4466
81. Hong S, Kim W (2015) Dynamics of water imbibition through paper channels with wax boundaries. Microfluid Nanofluidics 19:845–853
82. Masoodi R, Pillai KM (2012) Darcy's law-based model for wicking in paper-like swelling porous media. AICHE J 59:215–228
83. Perez-Cruz A, Stiharu I, Dominguez-Gonzalez A (2017) Two-dimensional model of imbibition into paper-based networks using Richards' equation. Microfluid Nanofluidics 21:1–12
84. Rolland JP, Mourey DA (2013) Paper as a novel material platform for devices. MRS Bull 38:299–305
85. Carrilho E, Martinez AW, Whitesides GM (2009) Understanding wax printing: a simple micropatterning process for paper-based microfluidics. Anal Chem 81:7091–7095
86. Zhang H, Smith E, Zhang W, Zhou A (2019) A. Inkjet printed microfluidic paper-based analytical device (μPAD) for glucose colorimetric detection in artificial urine. Biomed Microdevices 21:48
87. Kao PK, Hsu CC (2014) One-step rapid fabrication of paper-based microfluidic devices using fluorocarbon plasma polymerization. Microfluid Nanofluidics 16:811–818
88. Bruzewicz DA, Reches M, Whitesides GM (2008) Low-cost printing of poly(dimethylsiloxane) barriers to define microchannels in paper. Anal Chem 80:3387–3392
89. Osborn JL et al (2010) Microfluidics without pumps: reinventing the T-sensor and H-filter in paper networks. Lab Chip 10:2659–2665
90. Fu E, Ramsey SA, Kauffman P, Lutz B, Yager P (2011) Transport in two-dimensional paper networks. Microfluid Nanofluidics 10:29–35
91. Liu H, Crooks RM (2011) Three-dimensional paper microfluidic devices assembled using the principles of origami. J Am Chem Soc 133:17564–17566
92. Xu G et al (2016) Paper-Origami-based multiplexed malaria diagnostics from whole blood. Angew Chem Int Ed 55:15250–15253
93. Toley BJ et al (2013) Tunable-delay shunts for paper microfluidic devices. Anal Chem 85:11545–11552
94. Lutz B et al (2013) Dissolvable fluidic time delays for programming multi-step assays in instrument-free paper diagnostics. Lab Chip 13:2840–2847
95. Chen H, Cogswell J, Anagnostopoulos C, Faghri M (2012) A fluidic diode, valves, and a sequential-loading circuit fabricated on layered paper. Lab Chip 12:2909–2913
96. Yang X, Forouzan O, Brown TP, Shevkoplyas SS (2011) Integrated separation of blood plasma from whole blood for microfluidic paper-based analytical devices. Lab Chip 12: 274–280
97. Vella SJ et al (2012) Measuring markers of liver function using a micropatterned paper device designed for blood from a fingerstick. Anal Chem 84:2883–2891
98. Lutz B et al (2013) Dissolvable fluidic time delays for programming multi- step assays in instrument-free paper diagnostics. Lab Chip 13:2840–2847
99. Fu E et al (2012) Two-dimensional paper network format that enables simple multistep assays for use in low-resource settings in the context of malaria antigen detection. Anal Chem 84:4574–4579
100. Fridley GE, Le H, Yager P (2014) Highly sensitive immunoassay based on controlled rehydration of patterned reagents in a 2-dimensional paper network. Anal Chem 86:6447–6453

101. Achille C et al (2021) 3D printing of monolithic capillarity-driven microfluidic devices for diagnostics. Adv Mater 33:1–7
102. Hansson J, Yasuga H, Haraldsson T, Van Der Wijngaart W (2016) Synthetic microfluidic paper: high surface area and high porosity polymer micropillar arrays. Lab Chip 16: 298–304
103. de Puig H, Bosch I, Gehrke L, Hamad-Schifferli K (2017) Challenges of the nano–bio Interface in lateral flow and dipstick immunoassays. Trends Biotechnol 35:1169–1180
104. Lim H, Jafry AT, Lee J (2019) Fabrication, flow control, and applications of microfluidic paper-based analytical devices. Molecules 24: 1–32
105. Yamada K, Shibata H, Suzuki K, Citterio D (2017) Toward practical application of paper-based microfluidics for medical diagnostics: state-of-the-art and challenges. Lab Chip 17: 1206–1249
106. Wang X, Hagen JA, Papautsky I (2013) Paper pump for passive and programmable transport. Biomicrofluidics 7:1–11
107. Xu ZR et al (2008) A microfluidic flow injection system for DNA assay with fluids driven by an on-chip integrated pump based on capillary and evaporation effects. Lab Chip 8: 1658–1663
108. Wang J et al (2010) A self-powered, one-step chip for rapid, quantitative and multiplexed detection of proteins from pinpricks of whole blood. Lab Chip 10:3157–3162
109. Cummins BM, Chinthapatla R, Lenin B, Ligler FS, Walker GM (2017) Modular pumps as programmable hydraulic batteries for microfluidic devices. Technology 05:21–30
110. Kokalj T, Park Y, Vencelj M, Jenko M, Lee LP (2014) Self-powered imbibing microfluidic pump by liquid encapsulation: SIMPLE. Lab Chip 14:4329–4333
111. Dal Dosso F, Kokalj T, Belotserkovsky J, Spasic D, Lammertyn J (2018) Self-powered infusion microfluidic pump for ex vivo drug delivery. Biomed Microdevices 20:44
112. Vloemans D et al (2021) Precise sample metering method by coordinated burst action of hydrophobic burst valves applied to dried blood spot collection. Lab Chip 21:4445–4454
113. Dal Dosso F, Tripodi L, Spasic D, Kokalj T, Lammertyn J (2019) Innovative hydrophobic valve allows complex liquid manipulations in a self-powered channel-based microfluidic device. ACS Sensors 4:694–703
114. Dal Dosso F et al (2018) Creasensor: SIMPLE technology for creatinine detection in plasma. Anal Chim Acta 1000:191–198
115. McNeely MR, Sputea MK, Tusneem NA, Oliphant AR (1999) Sample processing with hydrophobic microfluidics. JALA J Assoc Lab Autom 4:30–33
116. Lim H et al (2022) Biosensors and bioelectronics smart bioelectronic pacifier for real-time continuous monitoring of salivary electrolytes. Biosens Bioelectron 210:114329
117. Koh A et al (2016) A soft, wearable microfluidic device for the capture, storage, and colorimetric sensing of sweat. Sci Transl Med 8:1–14
118. Park SM et al (2020) A mountable toilet system for personalized health monitoring via the analysis of excreta. Nat Biomed Eng 4: 624–635
119. Guthrie R, Susi A (1963) A simple phenylalanine method for detecting phenylketonuria in large populations of newborn infants. Pediatrics 32:338–343
120. Humbert P et al (2016) Skin capillaroscopy. In: Measuring the skin. https://doi.org/10.1007/978-3-319-26594-0
121. Bian S et al (2020) Evaluating an easy sampling method using dried blood spots to determine vedolizumab concentrations. J Pharm Biomed Anal 185:113224
122. Sciberras D et al (2019) A pharmacokinetic study of radiprodil oral suspension in healthy adults comparing conventional venous blood sampling with two microsampling techniques. Pharmacol Res Perspect 7:1–8
123. Karp DG et al (2020) A serological assay to detect SARS-CoV-2 antibodies in at-home collected finger-prick dried blood spots. Sci Rep 10:1–7
124. Popp W et al (2010) What is the use? An international look at reuse of single-use medical devices. Int J Hyg Environ Health 213: 302–307
125. Smit PW et al (2014) Systematic review of the use of dried blood spots for monitoring HIV viral load and for early infant diagnosis. https://doi.org/10.1371/journal.pone.0086461
126. Serafin A, Malinowski M, Prażmowska-Wilanowska A (2020) Blood volume and pain perception during finger prick capillary blood sampling: are all safety lancets equal? Postgrad Med 132:288–295
127. Morgan PE (2021) Microsampling devices for routine therapeutic drug monitoring-are we there yet? Ther Drug Monit 43:322–334

128. Chan PCR, Rozmanc M, Seiden-Long I, Kwan J (2009) Evaluation of a point-of-care glucose meter for general use in complex tertiary care facilities. Clin Biochem 42:1104–1112
129. Lenicek Krleza J, Dorotic A, Grzunov A, Maradin M (2015) Capillary blood sampling: national recommendations on behalf of the Croatian society of medical biochemistry and laboratory medicine. Biochem Med 25:335–358
130. Dungan K, Chapman J, Braithwaite SS, Buse J (2007) Glucose measurement: confounding issues in setting targets for inpatient management. Diabetes Care 30:403–409
131. Erbach M et al (2016) Interferences and limitations in blood glucose self-testing: an overview of the current knowledge. J Diabetes Sci Technol 10:1161–1168
132. Martial LC et al (2016) Cost evaluation of dried blood spot home sampling as compared to conventional sampling for therapeutic drug monitoring in children. PLoS One 11: e0167433
133. Denniff P, Spooner N (2010) The effect of hematocrit on assay bias when using DBS samples for the quantitative bioanalysis of drugs. Bioanalysis 2:1385–1395
134. Jones C, Dunseath GJ, Lemon J, Luzio SD (2018) Microsampling collection methods for measurement of C-peptide in whole blood. J Diabetes Sci Technol 12:1024–1028
135. Bloem K et al (2018) Capillary blood microsampling to determine serum biopharmaceutical concentration: mitra® microsampler vs dried blood spot. Bioanalysis 10:815–823
136. Veenhof H et al (2020) Volumetric absorptive microsampling and dried blood spot microsampling vs. conventional venous sampling for tacrolimus trough concentration monitoring. Clin Chem Lab Med 58:1687–1695
137. Leuthold LA et al (2015) New microfluidic-based sampling procedure for overcoming the hematocrit problem associated with dried blood spot analysis. Anal Chem 87:2068–2071
138. Lenk G et al (2015) A disposable sampling device to collect volume-measured DBS directly from a fingerprick onto DBS paper. Bioanalysis 7:2085–2094
139. Neto R, Gooley A, Breadmore MC, Hilder EF, Lapierre F (2018) Precise, accurate and user-independent blood collection system for dried blood spot sample preparation. Anal Bioanal Chem 410:3315–3323
140. Kim J-H, Woenker T, Adamec J, Regnier FE (2013) Simple, miniaturized blood plasma extraction method. https://doi.org/10.1021/ac402735y
141. Hauser A et al (2019) Evaluation of a BioRad avidity assay for identification of recent HIV-1 infections using dried serum or plasma spots. J Virol Methods 266:114–120
142. Tsai CW, Li CH, Lam RWK, Li CK, Ho S (2020) Diabetes care in motion: blood glucose estimation using wearable devices. IEEE Consum Electron Mag 9:30–34
143. Dixon RV et al (2021) Microneedle-based devices for point-of-care infectious disease diagnostics. Acta Pharm Sin B 11:2344–2361
144. Blicharz TM et al (2018) Microneedle-based device for the one-step painless collection of capillary blood samples. Nat Biomed Eng 2: 151–157
145. Li T, Barnett A, Rogers KL, Gianchandani YB (2009) A blood sampling microsystem for pharmacokinetic applications: design, fabrication, and initial results. Lab Chip. https://doi.org/10.1039/b910508e
146. Strambini LM et al (2020) Self-powered microneedle-based biosensors for pain-free high-accuracy measurement of glycaemia in interstitial fluid. Biosens Bioelectron 66: 162–168
147. Jiang X, Lillehoj PB (2020) Microneedle-based skin patch for blood-free rapid diagnostic testing. Microsyst Nanoeng 6:1–11
148. Liu L, Wang Y, Yao J, Yang C, Ding G (2016) A minimally invasive micro sampler for quantitative sampling with an ultrahigh-aspect-ratio microneedle and a PDMS actuator. Biomed Microdevices 18:59
149. Li CG, Lee CY, Lee K, Jung H (2013) An optimized hollow microneedle for minimally invasive blood extraction. Biomed Microdevices 15:17–25
150. Miller PR et al (2014) Microneedle-based transdermal sensor for on-chip potentiometric determination of K+. Adv Healthc Mater 3:876–881
151. Evens T et al (2021) Producing hollow polymer microneedles using laser ablated molds in an injection molding process. J Micro Nano-Manufacturing 9:1–9
152. Chaudhri BP, Ceyssens F, De Moor P, Van Hoof C, Puers R (2010) A high aspect ratio SU-8 fabrication technique for hollow microneedles for transdermal drug delivery and blood extraction. J Micromech Microeng 20:064006

153. Kashaninejad N et al (2021) Microneedle arrays for sampling and sensing skin interstitial fluid. Chemosensors 9:83
154. Miller PR et al (2018) Extraction and biomolecular analysis of dermal interstitial fluid collected with hollow microneedles. Commun Biol 1:173
155. Kolluru C et al (2019) Monitoring drug pharmacokinetics and immunologic biomarkers in dermal interstitial fluid using a microneedle patch. Biomed Microdevices. https://doi.org/10.1007/s10544-019-0363-3
156. Romanyuk AV et al (2014) Collection of analytes from microneedle patches. Anal Chem 86:10520–10523
157. Lee H et al (2019) Porous microneedle integrated in paper based glucose sensor for fluid channel interface. In: 2019 IEEE CPMT symposium Japan, ICSJ 2019, pp 39–42
158. Takeuchi K et al (2022) Microfluidic chip connected to porous microneedle array for continuous ISF sampling. Drug Deliv Transl Res 12:435–443
159. Li CG et al (2015) One-touch-activated blood multidiagnostic system using a minimally invasive hollow microneedle integrated with a paper-based sensor. Lab on a Chip 15(16) 3286–3292. https://doi.org/10.1039/C5LC00669D
160. Rifai N, Gillette MA, Carr SA (2006) Protein biomarker discovery and validation: the long and uncertain path to clinical utility. Nat Biotechnol 24:971–983
161. Wide L, Gemzell CA (1960) An immunological pregnancy test. Acta Endocrinol 35:261–267
162. Scheiblauer H et al (2021) Comparative sensitivity evaluation for 122 CE-marked rapid diagnostic tests for SARS-CoV-2 antigen, Germany, September 2020 to April 2021. Eur Secur 26:2100441
163. Gray ER et al (2018) A landscape review of antigen detection for early HIV diagnosis. AIDS 32:2089–2102
164. Mortimer PP, Parry JV (1994) Detection of antibody to HIV in saliva: a brief review. Clin Diagn Virol 2:231–243
165. Tian T, Bi Y, Xu X, Zhu Z, Yang C (2018) Integrated paper-based microfluidic devices for point-of-care testing. Anal Methods 10: 3567–3581
166. Kim D, Kim S, Kim S (2020) An innovative blood plasma separation method for a paper-based analytical device using chitosan functionalization. Analyst 145:5491–5499
167. Nilghaz A, Shen W (2015) Low-cost blood plasma separation method using salt functionalized paper. RSC Adv 5:53172–53179
168. Bhamla MS et al (2017) Hand-powered ultra-low-cost paper centrifuge. Nat Biomed Eng 1:1–7
169. Songjaroen T, Dungchai W, Chailapakul O, Henry CS, Laiwattanapaisal W (2012) Blood separation on microfluidic paper-based analytical devices. Lab Chip 12:3392–3398
170. Zhang H et al (2021) A low-cost mobile platform for whole blood glucose monitoring using colorimetric method. Microchem J 162:105814
171. Choobbari ML, Rad MB, Jahanshahi A, Ghourchian H (2020) A sample volume independent paper microfluidic device for quantifying glucose in real human plasma. Microfluid Nanofluidics 24:74
172. Tenda K et al (2018) Paper-based antibody detection devices using bioluminescent BRET-switching sensor proteins. Angew Chem Int Ed 57:15369–15373
173. Lutz BR, Trinh P, Ball C, Fu E, Yager P (2011) Two-dimensional paper networks: programmable fluidic disconnects for multi-step processes in shaped paper. Lab Chip 11: 4274–4278
174. Tu D, Holderby A, Dean J, Mabbott S, Coté GL (2021) Paper microfluidic device with a horizontal motion valve and a localized delay for automatic control of a multistep assay. Anal Chem 93:4497–4505
175. Fu H et al (2019) A paper-based microfluidic platform with shape-memory-polymer-actuated fluid valves for automated multi-step immunoassays. Microsyst Nanoeng 5:1–12
176. Hui CY, Liu M, Li Y, Brennan JD (2018) A paper sensor printed with multifunctional bio/nano materials. Angew Chem Int Ed 57:4549–4553
177. Liu X et al (2022) A paper-based all-in-one origami nanobiosensor for point-of-care diagnosis of cardiovascular diseases. https://papers.ssrn.com/abstract=4060674
178. Chen C-A et al (2021) An electricity- and instrument-free infectious disease sensor based on a 3D origami paper-based analytical device. Lab Chip 21:1908–1915
179. Lee M-J et al (2022) Pumpless three-dimensional photo paper–based microfluidic analytical device for automatic detection of thioredoxin-1 using enzyme-linked immunosorbent assay. Anal Bioanal Chem 414:3219–3230
180. Deraney RN, Mace CR, Rolland JP, Schonhorn JE (2016) Multiplexed, patterned-paper immunoassay for detection of malaria and dengue fever. Anal Chem 88:6161–6165
181. Hu SW et al (2017) Dual-functional carbon dots pattern on paper chips for Fe3+ and

ferritin analysis in whole blood. Anal Chem 89:2131–2137
182. Park J, Park JK (2017) Pressed region integrated 3D paper-based microfluidic device that enables vertical flow multistep assays for the detection of C-reactive protein based on programmed reagent loading. Sensors Actuators B Chem 246:1049–1055
183. Mu X et al (2015) A paper-based skin patch for the diagnostic screening of cystic fibrosis. Chem Commun 51:6365–6368
184. Yamada K, Henares TG, Suzuki K, Citterio D (2015) Distance-based tear Lactoferrin assay on microfluidic paper device using interfacial interactions on surface-modified cellulose. ACS Appl Mater Interfaces 7:24864–24875
185. Cate DM, Dungchai W, Cunningham JC, Volckens J, Henry CS (2013) Simple, distance-based measurement for paper analytical devices. Lab Chip 13:2397–2404
186. Tian T et al (2016) Integration of target responsive hydrogel with cascaded enzymatic reactions and microfluidic paper-based analytic devices (μPADs) for point-of-care testing (POCT). Biosens Bioelectron 77:537–542
187. Li M, Tian J, Al-Tamimi M, Shen W (2012) Paper-based blood typing device that reports patient's blood type 'in writing'. Angew Chem Int Ed Engl 51:5497–5501
188. Lan T, Zhang J, Lu Y (2016) Transforming the blood glucose meter into a general healthcare meter for in vitro diagnostics in mobile health. Biotechnol Adv 34:331–341
189. Ghosh S et al (2020) A new microchannel capillary flow assay (MCFA) platform with lyophilized chemiluminescence reagents for a smartphone-based POCT detecting malaria. Microsyst Nanoeng 6:5
190. Tsao YT et al (2021) Point-of-care semen analysis of patients with infertility via smartphone and colorimetric paper-based diagnostic device. Bioeng Transl Med 6:1–11
191. Tian T et al (2017) Integrated distance-based origami paper analytical device for one-step visualized analysis. ACS Appl Mater Interfaces 9:30480–30487
192. Achille C et al (2021) Microfluidic devices: 3D printing of monolithic capillarity-driven microfluidic devices for diagnostics (Adv. Mater. 25/2021). Adv Mater 33:2170192
193. Zimmermann M, Hunziker P, Delamarche E (2009) Autonomous capillary system for one-step immunoassays. Biomed Microdevices 11:1–8
194. Hemmig E, Temiz Y, Gökçe O, Lovchik RD, Delamarche E (2020) Transposing lateral flow immunoassays to capillary-driven microfluidics using self-coalescence modules and capillary-assembled receptor carriers. Anal Chem 92:940–946
195. Yafia M et al (2022) Microfluidic chain reaction of structurally programmed capillary flow events. Nature 605:464–469
196. Qu JH et al (2022) Point-of-care therapeutic drug monitoring of adalimumab by integrating a FO-SPR biosensor in a self-powered microfluidic cartridge. Biosens Bioelectron 206:114125
197. Ordutowski H et al (2022) Next generation point-of-care test for therapeutic drug monitoring of adalimumab in patients diagnosed with autoimmune diseases. Biosens Bioelectron 208:114189
198. Mullis K et al (1986) Specific enzymatic amplification of DNA in vitro: the polymerase chain reaction. Cold Spring Harb Symp Quant Biol 51:263–273
199. Okeke IN, Ihekweazu C (2021) The importance of molecular diagnostics for infectious diseases in low-resource settings. Nat Rev Microbiol 19:547–548
200. Elshafie S, Taj-Aldeen SJ (2016) Emerging resistant serotypes of invasive Streptococcus pneumoniae. Infect Drug Resist 9:153–160
201. Okeke IN et al (2020) Leapfrogging laboratories: the promise and pitfalls of high-tech solutions for antimicrobial resistance surveillance in low-income settings. BMJ Glob Health 5:1–9
202. Mauk MG, Song J, Tong Y, Bau HH, Liu C (2015) Translating nucleic acid amplification assays to the microscale: lab on a chip for point-of-care molecular diagnostics. https://doi.org/10.2174/157341101266151020212121
203. Zhang JY, Bender AT, Boyle DS, Drain PK, Posner JD (2021) Current state of commercial point-of-care nucleic acid tests for infectious diseases. Analyst 146:2449–2462
204. Saeed M et al (2017) GeneXpert technology: a breakthrough for the diagnosis of tuberculous pericarditis and pleuritis in less than 2 hours. Saudi Med J 38:699–705
205. Saeed M et al (2017) Genexpert: a new tool for the rapid detection of rifampicin resistance in mycobacterium tuberculosis. J Pak Med Assoc 67:270–274
206. Huang H et al (2019) Evaluation, validation, and implementation of the Idylla system as rapid molecular testing for precision medicine. J Mol Diagn 21:862–872
207. Poritz MA et al (2011) Filmarray, an automated nested multiplex PCR system for multi-pathogen detection: development and

application to respiratory tract infection. PLoS One 6:e26047
208. Nie S et al (2014) Evaluation of alere i influenza a&B for rapid detection of influenza viruses a and B. J Clin Microbiol 52:3339–3344
209. Chen D et al (2010) An integrated, self-contained microfluidic cassette for isolation, amplification, and detection of nucleic acids. Biomed Microdevices 12:705–719
210. Craw P, Balachandran W (2012) Isothermal nucleic acid amplification technologies for point-of-care diagnostics: a critical review. Lab Chip 12:2469–2486
211. De Paz HD, Brotons P, Munoz-Almagro C (2014) Molecular isothermal techniques for combating infectious diseases: towards low-cost point-of-care diagnostics. Expert Rev Mol Diagn 14:827–844
212. Schrader C, Schielke A, Ellerbroek L, Johne R (2012) PCR inhibitors – occurrence, properties and removal. J Appl Microbiol 113:1014–1026
213. Li J, Macdonald J (2014) Advances in isothermal amplification: novel strategies inspired by biological processes. Biosens Bioelectron 64: 196–211
214. Fakruddin M et al (2013) Nucleic acid amplification: alternative methods of polymerase chain reaction. J Pharm Bioallied Sci 5:245–252
215. Obande GA, Singh KKB (2020) Current and future perspectives on isothermal nucleic acid amplification technologies for diagnosing infections. Infect Drug Resist 13:455–483
216. Fronczek CF, Park TS, Harshman DK, Nicolini AM, Yoon JY (2014) Paper microfluidic extraction and direct smartphone-based identification of pathogenic nucleic acids from field and clinical samples. RSC Adv 4: 11103–11110
217. Govindarajan AV, Ramachandran S, Vigil GD, Yager P, Böhringer KF (2012) A low cost point-of-care viscous sample preparation device for molecular diagnosis in the developing world; an example of microfluidic origami. Lab Chip 12:174–181
218. Rodriguez NM et al (2015) Paper-based RNA extraction, in situ isothermal amplification, and lateral flow detection for low-cost, rapid diagnosis of influenza a (H1N1) from clinical specimens. Anal Chem 87:7872–7879
219. Rodriguez NM, Wong WS, Liu L, Dewar R, Klapperich CM (2016) A fully integrated paperfluidic molecular diagnostic chip for the extraction, amplification, and detection of nucleic acids from clinical samples. Lab Chip 16:753–763
220. Rohrman BA, Richards-Kortum RR (2012) A paper and plastic device for performing recombinase polymerase amplification of HIV DNA. Lab Chip 12:3082–3088
221. Cordray MS, Richards-Kortum RR (2015) A paper and plastic device for the combined isothermal amplification and lateral flow detection of plasmodium DNA. Malar J 14: 1–8
222. Connelly JT, Rolland JP, Whitesides GM (2015) 'Paper machine' for molecular diagnostics. Anal Chem 87:7595–7601
223. Reboud J et al (2019) Paper-based microfluidics for DNA diagnostics of malaria in low resource underserved rural communities. Proc Natl Acad Sci USA 116:4834–4842
224. Caliendo AM et al (2013) Better tests, better care: improved diagnostics for infectious diseases. Clin Infect Dis 57(Suppl 3): 139–170
225. Lafleur LK et al (2016) A rapid, instrument-free, sample-to-result nucleic acid amplification test. Lab Chip 16:3777–3787
226. Tang R et al (2017) A fully disposable and integrated paper-based device for nucleic acid extraction, amplification and detection. Lab Chip 17:1270–1279
227. Yeh E-C et al (2017) Self-powered integrated microfluidic point-of-care low-cost enabling (SIMPLE) chip. Sci Adv 3:1–11
228. Buser JR et al (2015) Precision chemical heating for diagnostic devices. Lab Chip 15: 4423–4432
229. Liao S-C et al (2016) Smart cup: a minimally-instrumented, smartphone-based point-of-care molecular diagnostic device. Sensors Actuators B Chem 229:232–238
230. Singleton J et al (2013) Instrument-free exothermic heating with phase change temperature control for paper microfluidic devices. SPIE 86:3279–3288
231. Goertz JP et al (2018) Multistage chemical heating for instrument-free biosensing. ACS Appl Mater Interfaces 10:33043–33048
232. Huang S et al (2013) Low cost extraction and isothermal amplification of DNA for infectious diarrhea diagnosis. PLoS One 8:1–10
233. Liu C, Mauk MG, Hart R, Qiu X, Bau HH (2011) A self-heating cartridge for molecular diagnostics. Lab Chip 11:2686–2692
234. Vloemans D, Dal Dosso F, Verboven P, Nicolai B, Lammertyn J (2020) Exploiting phase change materials in tunable passive heating system for low-resource point-of-care diagnostics. Appl Therm Eng 173: 115269

Part II

Microchips for Sample Preparation Including Cell-Free DNA, Exosomes, and Cells

Chapter 2

DNA Sizing with pg/μL Sensitivity for Cell-Free DNA Analysis with the μLAS Technology

Audrey Boutonnet, Inga Tijunelyte, Aurélien Bancaud, and Frédéric Ginot

Abstract

The μLAS technology enables in-line DNA concentration and separation in a microchannel. Here, we describe its operation to analyze the size profile of cell-free DNA (cfDNA) extracted from blood plasma. Operated on commercial systems for capillary electrophoresis, we provide the size distribution of healthy individuals or patients using an input of 10 μL.

Key words DNA, cfDNA, Sizing, Microfluidics, Capillary, Separation

1 Introduction

Cancer diagnostic and follow-up are essentially carried out with radiology techniques, tissue biopsy analysis, and body fluid analysis. While the first two -methods can lead to several adverse effects [1, 2], notably associated with surgical operations or exposure to irradiations, the detection of biomarkers in body fluids is minimally invasive. Among these biomarkers, there is an upsurge in studies considering cell-free DNA in body fluids (cfDNA) to stratify patients, monitor the treatment response, and predict disease progression [3]. The characterization of cfDNA nevertheless remains a challenge due to its low concentration which typically spans 1 to 100 ng/mL, and broad size distribution from 100 to 5000 bp. It can either be performed with target gene analysis, using enzymatic amplification to achieve detection [1], and/or by global size analysis using high-sensitivity DNA separation technologies. The latter technologies generally include a concentration step so as to obtain a detectable signal by fluorescence microscopy. Among the different technological solutions [4–6], we describe the μLAS technology in microfluidic format and its version for capillary electrophoresis,

Valérie Taly et al. (eds.), *Microfluidics Diagnostics: Methods and Protocols*, Methods in Molecular Biology, vol. 2804, https://doi.org/10.1007/978-1-0716-3850-7_2,

termed the BIABooster, that performs sizing and concentration of double-stranded DNA with a sensitivity of 10 fg/μL and provides the size distribution of cfDNA analysis in minutes.

2 Materials

2.1 μLAS Using Microfluidic Chips

1. Microfluidic chips: μLAS chips are fabricated by photolithography, reactive ion etching, thermal growth of a 300 nm layer of silicon oxide, and silicon-glass anodic bonding (see detailed protocol in [7]). The microchannel consists of a linear channel with a contraction midway through the inlet and outlet (see Fig. 1a for dimensions). The channel height for the linear part and the constriction are 16 and 2 μm, respectively. The constriction of the chip is designed with a linear head-to-head with a power-law geometry (Fig. 1b). The user chooses the use of single- or double-channel chips (Fig. 1c).
2. Supports to hold microfluidic chips containing inlet and outlet reservoirs sealed by O-rings and 0.5 mm diameter holes pierced sidewalls for integrating platinum wire electrodes (Fig. 2a).
3. Pressure controller (7 bar, Fluigent) with pneumatic fittings.
4. High voltage power supply (Labsmith).
5. Epifluorescence microscope equipped with a light engine for cyan excitation (Lumencore) and an sCMOS digital camera (Andor) as well as 20× or 10× air objectives.
6. Ethanol.

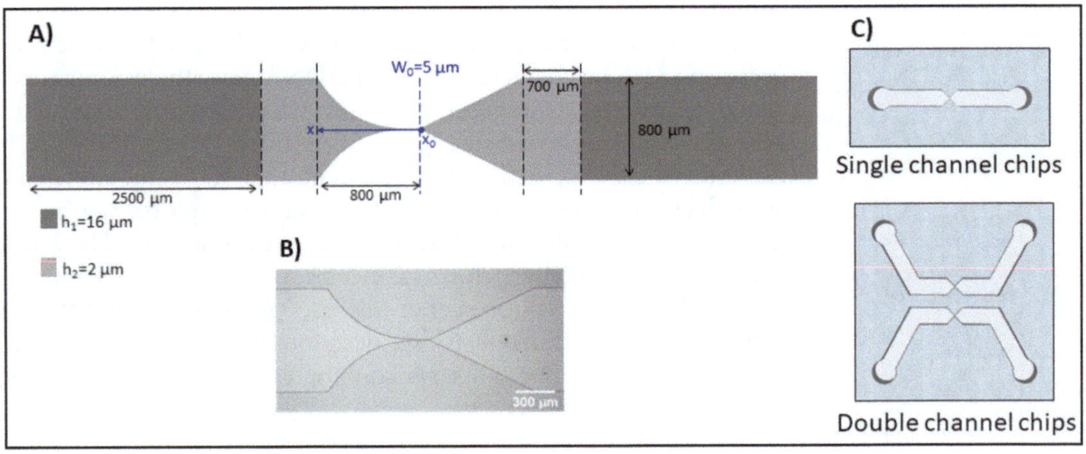

Fig. 1 (**a**) Design and dimensions of μLAS chip. (**b**) Bright-field image of the funnel region. (**c**) Design of the single or double channel chips

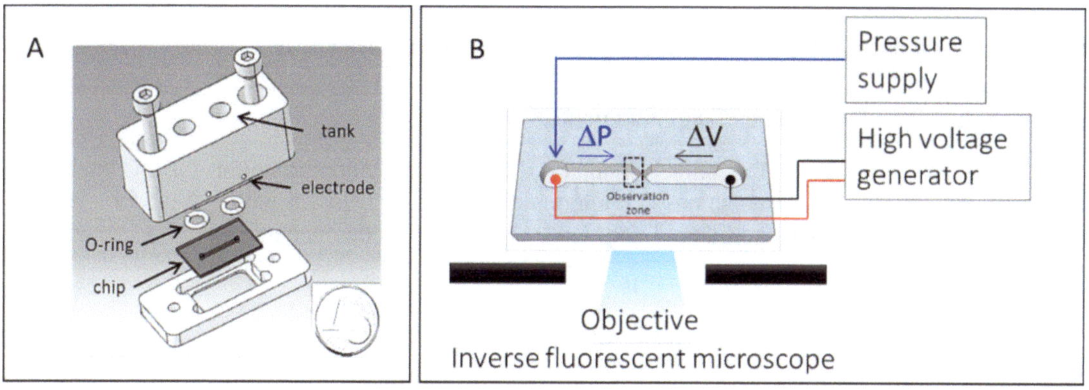

Fig. 2 (**a**) Microfluidic chip integration for its operation. (**b**) Strategy for electrohydrodynamic actuation and observation by fluorescence microscopy. (Image extracted from [8])

7. PVP-based working buffer: Prepare solutions with ultrapure water (18 MΩ-cm at 25 °C) and store them at room temperature (RT). Dissolve neutral polymer polyvinylpyrrolidone (PVP; MW = 1.3 MDa) at a 5% (w:v) concentration in electrolyte solution such as TBE 1× (*see* **Note 1**). Perform vortex mixing of the solution for 2 min. Incubate the solution on a rotator overnight at RT (*see* **Note 2**). Prior to use, filter the working buffer with 0.2 μm polyethersulfone (PES) filters.
8. DNA ladders: Dilute the DNA ladder in the working buffer. DNA concentrations on the order of 10–7 M ensure fast visualization of DNA bands. Vortex the mixture for 2 min. Leave the solution at room temperature for 1 h before use.

2.2 μLAS Using the BIABooster Device in a Capillary Electrophoresis Instrument

1. CE-LIF System: Agilent Technologies Capillary Electrophoresis 7100 system equipped with an Adelis Zetalif LASER488 or a Zetalif LED480 fluorescence detector (*see* **Notes 3, 4** and **5**). The capillary is a BIABooster device (Adelis). An external pressure system (compressed air or nitrogen) at 12 bar is necessary. The pressure and electric field are applied at the same time. We use the CEC mode, as selected in CE configuration in OpenLAB Chemstation (Agilent Technologies).
2. Hydrochloric acid solution (0.1 M): Mix 833 μL of concentrated hydrochloric acid in 99.2 mL of water (molecular biology grade water) using a magnetic stirrer for 5 min.
3. BIABooster Kit for DNA Sizing (DNA 1 K Kit, Adelis) including Capillary washing solution, capillary conditioning solution (ref: 16-BB-DNA1K/09), running buffer (Running buffer DNA 1 K solution, ref.: 16-BB-DNA1K/10; *see* **Note 6** for home-made running buffer), fluorescent dye (ref:16-BB-DNA1K/02; *see* **Note 7** for alternative dyes), DNA ladder 1 K, 5 mL syringe and syringe filter.
4. RNAse1, 10 U/μL (ThermoFisher).

Table 1
Number of vials used in the CE carrousel for different numbers of analysis

Required number of vials depending on number of samples								
Number of samples	**HCl 0.1 M**	**water**	**Capillary washing solution**	**Capillary conditioning solution**	**Running buffer DNA 1 K**	**Running buffer DNA 1 K with fluorescent dye**	**DNA ladder 1 K**	**Waste vials**
From 1 to 5	2	3	1	1	1	2	1	3
From 1 to 10	2	3	2	1	2	4	2	5
From 1 to 15	3	3	2	2	3	6	3	6

5. Running buffer with dye: Pipette 2 mL of running buffer in a tube. Add 20 μL of fluorescent dye; Vortex for 10 s. Collect this solution with a 5 mL syringe. Place a syringe filter on the syringe and filter the solution in two polypropylene vials. *See* **Note 8**.

2.3 Required Vials for Sample Analysis

Table 1 indicates the number of vials required depending on the number of analyzed samples.

1. Standard glass vials (Agilent Technologies, P/N: 5182-9697) for dispensing liquid reagents and solutions.
2. Polypropylene vials (Agilent Technologies, P/N: 5182-0567) to reduce dye adsorption during mixing running buffer with dye.
3. Glass inserts (Adelis, P/N:16-BB-DNA1K/06, volume 10 μL) to reduce the required volume during DNA ladder preparation.

2.4 Cell-Free DNA Analysis

1. Blood tube samples for plasma preparation.
2. Centrifuge.
3. DNA extraction kit: QIAmp Circulating Nucleic Acid Kit (QIAGEN), IDXTRACT kit (ID-Solutions), or Maxwell® RSC cfDNA Plasma Kits (Promega).

3 Methods

3.1 Operation with Microfluidic Chips: Experimental Setup

1. Place the microfluidic chips into a support.
2. Close the tanks with pneumatic fittings to connect the chip to the pressure controller.
3. For electric field actuation, connect the high-voltage power supply to the electrodes. Make sure that the electric force is oriented opposite to the flow (Fig. 2b).
4. Place the chip with its support on the epifluorescence microscope for visualization (*see* **Note 9**).

3.2 Operation with Microfluidic Chips: Experimental Flow Data Acquisition

1. Before the experiment, fill microfluidic chips with ethanol using 0.5 bar pressure, and flush for 10 min to remove bubbles.
2. Rinse the chip with milli-Q water for 10 min at 0.8 bar.
3. Inject the working buffer with a pressure of 1.2 bar and flush for 10 min to ensure surface coating with the PVP.
4. Introduce the solution containing labeled DNA and switch the microscope to fluorescence mode. Extensive rinsing with working buffer is needed between each analysis (*see* **Note 10**).
5. Apply constant pressure drop. Pressure can vary between 2 and 7 bar (see example of analysis).
6. Apply voltage in the range of 50–400 V (see example of analysis, *see* **Note 11**).
7. The concentration and separation of DNA bands are observed immediately. User can choose to take a snapshot or to record a movie of concentration and separation.
8. Post-proceeding involves extracting the fluorescence intensity along the symmetry line of the funnel (blue and red arrows in Fig. 3a and *x* axis in Fig. 3b). Intensity profiles are then fit using the multipeak Gaussian functions to extract the position and width of each peak.

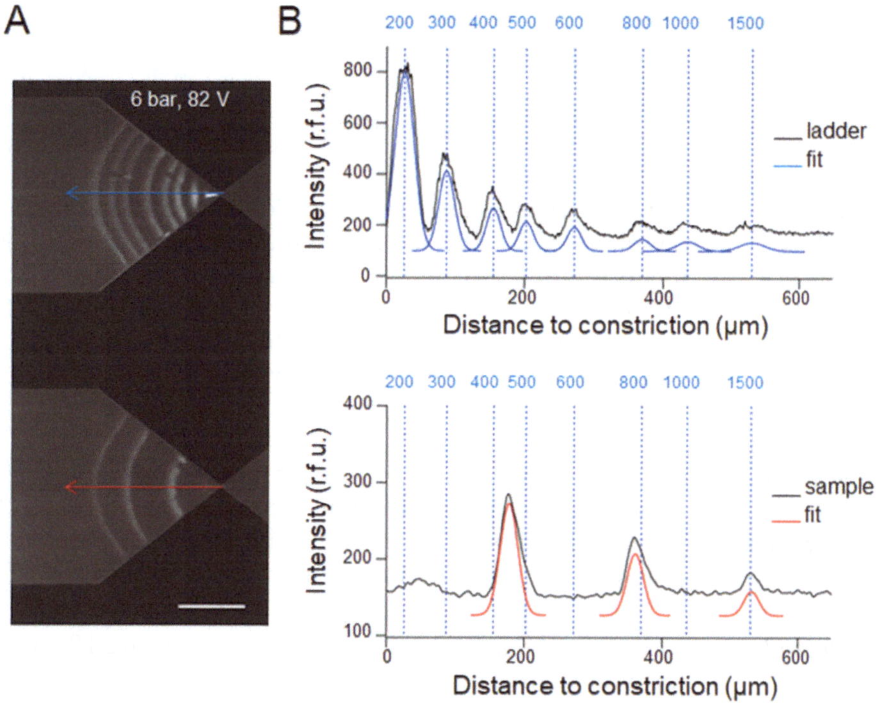

Fig. 3 (**a**) A fluorescence micrograph with the two-channel chips after a concentration time of 30 s. (**b**) The fluorescence intensity profiles corresponding to each micrograph are shown in black and the corresponding fits using Gaussian functions are in blue and red. The scale bars correspond to 300 μm. (Adapted from [7])

Table 2
Conditioning BIABooster device

Inlet vial	Outlet vial	Pressure (Bar)	Voltage (kV)	Duration (min)
Waste	Capillary washing solution	−10	0	3
Capillary washing solution	Waste	10	0	1.50
HCl 0.1 M	Waste	10	0	2
Capillary conditioning solution	Waste	10	0	4
Running buffer with dye	Waste	10	0	6

Table 3
Sample injection

Inlet vial	Outlet vial	Pressure (Bar)	Voltage (kV)	Duration (s)
Sample	Waste	5	0	80
Water	Waste	Load inlet	0	0
Water	Waste	Load inlet	0	0
Running buffer with dye	Waste	5	0	10
Water	Waste	Load inlet	0	0
Water	Waste	Load inlet	0	0

3.3 Operation with Microfluidic Chips: DNA Separation in a Microfluidic Chip

A double-channel chip is used to identify DNA sizes in an unknown sample (Fig. 3). The analytical sample consists of three DNA fragments of 466, 798, and 1512 bp mixed at the same concentration of 80 pg/μL. The reference channel contains a 100 bp DNA ladder at the concentration of 500 ng/μL. We used 6 bar and 82 V, corresponding to a maximum flow velocity and an electric field of 7 cm/s and 5 kV/cm, respectively.

3.4 Operation with the BIABooster System: Experimental Setup

The analytical protocol consists of three steps characterized by the following electrohydrodynamic actuation parameters: (1) conditioning BIABooster device (Table 2); (2) sample injection (Table 3); and (3) DNA concentration and separation (Table 4).

3.5 Operation with the BIABooster System: DNA Separation in Capillaries

In Fig. 4, a DNA1K ladder is concentrated using 10 bar and 25 kV for 6 min. After this concentration step, the electric field is gradually reduced from 25 to 0.1 kV in order to separate DNA. The limit of detection (signal-to-noise ratio equal to 3) of the bands of 100, 200, 300, 400, and 1000 bp are 0.3, 0.07, 0.04, 0.02, and

Table 4
DNA concentration and separation

Inlet	Outlet	Pression (Bar)	Voltage (kV)	Duration (min)
Running buffer with dye	Running buffer with dye	10	25	0
		10	25	6
		10	10	6.4
		10	5	6.8
		10	0.6	10
		10	0.1	21
		10	0	21.1

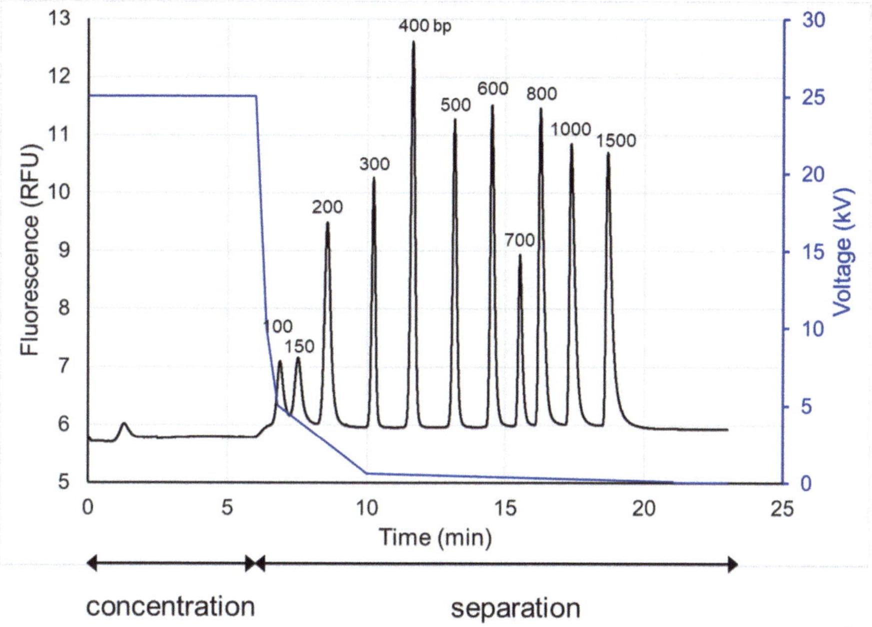

Fig. 4 BIABooster analysis of DNA1K ladder (0.1–1.5 kb) at a concentration of 32 pg/μL. The black and blue curves correspond to the fluorescence signal and the applied voltage during analysis, respectively. Above each peak, the fragment size is reported in bp

0.01 pg/μL, respectively. To determine the size profile and concentration of unknown samples, a DNA1K ladder is processed every five samples.

3.6 Operation with the BIABooster System: Analysis of cfDNA

We describe a plasma preparation method suitable for cfDNA analysis based on Ref. [9] (*see* **Note 12**).

3.6.1 Plasma Preparation and cfDNA Extraction

1. Centrifuge blood tubes at 1200 g for 10 min at room temperature.
2. Transfer supernatant to a new tube, avoiding pipetting the buffy coat.

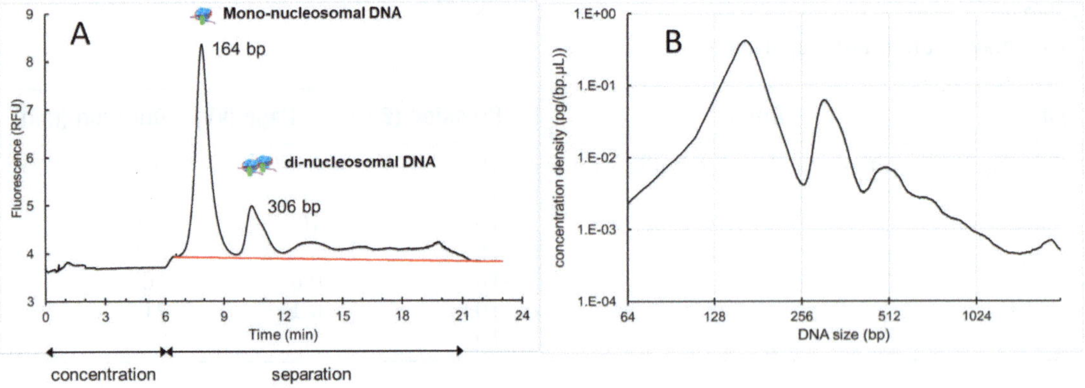

Fig. 5 Typical cfDNA size profile. (**a**) The black curve corresponds is the fluorescent trace with two main peaks the size of which matches that of mono and di-nucleosomes. The red line is the base line. (**b**) The fluorescence intensity plot of A is converted into a concentration per bp over size

3. Centrifuge samples at 16,000 g for 10 min at 4 °C (*see* **Note 13**).
4. Proceed to cfDNA extraction, or store the sample at −80 °C (*see* **Note 14**). For this step, use your preferred extraction method provided that the QIAmp Circulating Nucleic Acid Kit (*see* **Note 15**), the IDXTRACT kit, and the Maxwell® RSC cfDNA Plasma Kits have been tested successfully (*see* **Note 16**).

3.6.2 RNAse Treatment (See **Notes 17** and **18**)

1. Dilute the RNAseI solution at 0.2 U/µL in TE 1×.
2. Add 5 µL of this solution in 5 µL of cfDNA sample.
3. Incubate at 37 °C for 30 min.

3.6.3 BIABooster Analysis

1. Proceed as described above for DNA samples (*see* **Note 19**). A typical chromatogram is reported in Fig. 5a, and its conversion into a size distribution in Fig. 5b.
2. BIABooster system is relevant to control the quality of cfDNA samples. Figure 6a shows a raw profile of a cfDNA sample containing excessive amounts of high molecular weight DNA. Figure 6b shows a degraded cfDNA sample, as inferred from the fluorescence signal at the end of the concentration step, between 3 and 6 min, which indicates the presence of molecules smaller than 75 bp.

3.6.4 Data Treatment

Conventional data treatment designed for chromatographic or electrophoresis systems, as described in [10], must be adapted for BIABooster analyses, because the migration of DNA fragments is not constant over time.

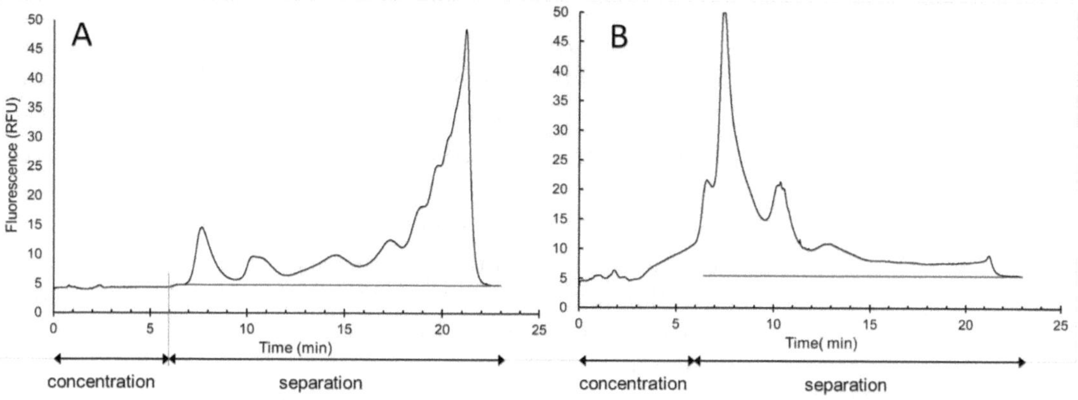

Fig. 6 Anomalous cfDNA profiles. (**a**) The profile contains a high amount of high molecular weight DNA. It is typically obtained with blood samples collected in EDTA tubes, after a few days of aging at RT before plasma preparation. (**b**) cfDNA profile containing a lot of small DNA (<75 bp)

1. Export raw data from the Agilent CE in a CSV format using the Export function:
   (a) In Data Analysis, select Export function.
   (b) Complete the parameters: export directory, unit signal.
   (c) Select the sequence file to export.
   (d) Files in CSV format are then saved in the export folder (*see* **Note 20**).
2. Import data into BIABooster Analytics software and perform analysis (or perform data treatment with other means, *see* **Note 21**).

4 Notes

1. Different electrolytes can be selected depending on the application. The conductivity as well as to a lesser extent the viscosity depends on the electrolyte. If you change electrolyte, adjust the electric field for the concentration step.
2. There are different ways to prepare a PVP-based working buffer, including leaving it overnight at RT on a rotator, or stir it for 2 h at 40 °C on a magnetic stirrer. The viscosity of the resulting solution is largely dependent on the chosen method. We advise to prepare the running buffer with the same method for comparing experiments.
3. The CE-LIF system simultaneously delivers high pressure (up to 10 bar) and high voltage (25 kV). If you have a CE-LIF equipment in your lab, Adelis can study a specific adaptation of the BIABooster device to your CE-LIF system.

4. For this application, the fluorescence detectors Zetalif LASER 488 nM and Zetalif LED 480 nM are equivalent. The limit of detection is due to the background signal in the running buffer, which contains the dye.
5. A short electrode is used on the inlet side in order to avoid any contamination of the sample. The short electrode (Agilent Technologies) is 5 mm shorter than the standard.
6. For a home-made buffer, you may use TBE 1× with PVP, as for microfluidic chips. It is preferable to use TBE 0.5× to decrease electric current, hence Joule heating, when applying 25 kV or more.
7. You may also use YOPRO-1, YOYO-1, or Sybr Green (Molecular Probes, Invitrogen). The dye/DNA base pair ratio is 1: 8 for YOYO-1 and YOPRO-1.
8. The filtration step removes fluorescent microaggregates, which generate small spikes in the fluorescence traces. The running buffer + dye solution is stable for 1 day at room temperature in the carrousel of the CE.
9. The excitation wavelength is chosen regarding the fluorescence dye used to label DNA. The dyes used in our studies (YOYO-1 and YO-PRO-1) are excited at 491 nM and emit at 509 nM.
10. After conditioning the chip with a PVP-based working buffer, make sure to not introduce ethanol in between analyses. Otherwise, it leads to precipitates that immediately clog the constriction and irreversibly damage the chip. If an air bubble arises during the experiment, try to remove it by applying the same high pressure to both reservoirs.
11. A 300 nm layer of silicon oxide silicon is grown on silicon surfaces to avoid breakdown at high electric fields. We suggest to operate at tensions lower than 300–400 V.
12. Any plasma preparation suitable for cfDNA analysis by PCR or sequencing will work. The prerequisites are (i) the preparation of plasma within the first 6 h after blood drawing using standard K2-EDTA collecting tubes, and (ii) the use of a double centrifugation to completely eliminate cells and cell debris. Tubes specifically designed for cfDNA analysis (Cell-free Collection Tube (Roche Diagnostics), PAXgene Blood ccfDNA tube (Qiagen), and Streck BCT tubes (Streck)), which allow keeping the blood at room temperature for several days before preparing plasma, have been used successfully for cfDNA sizing.
13. Centrifuging 10 mL tubes at 16,000 g at 4 °C requires an appropriate centrifuge. We also ran the second centrifugation at 4500 g at room temperature. The resulting plasma is less clear when a turbid plasma is obtained after the first

centrifugation. However, cfDNA size profiles are reliable. In any case, use exactly the same preanalytical method for all the samples of a given study.

14. In order to avoid multiple freezing/defreezing cycles, it is recommended to aliquot plasma samples before freezing. In our hands, freezing/defreezing plasma samples one time has no impact on cfDNA size distribution, though it may impact the total concentration.
15. This analytical method is suitable for unsalted samples. The concentration step can be adapted to perform cfDNA sizing without prior extraction, starting from 100 μL of plasma.
16. We noticed differences in the size distribution of cfDNA according to the DNA extraction method. cfDNA kits have generally been set up for small fragments extraction, and the yield for high MW molecules varies significantly. For instance, the concentration of long molecules (>500 bp) is lower with the IDXtract kit than the QIAMP CNA kit. Also, the size of the main mononucleosomal peak slightly could change from one kit to another, e.g., the peak is 3 bp smaller with the QIAMP CNA kit than the IDXtract kit. The main features of the size profile are nevertheless conserved between the different procedures.
17. When using a DNA extraction procedure with a carrier to improve extraction yield, note that the carrier is likely to interfere with the size analysis, especially if it contains nucleic acids. For instance, the carrier in the Qiagen CNA kit contains RNA. We recommend not to use the carrier given that we repeatedly compared the yield of cfDNA extraction with and without carrier, and never detected any difference, in agreement with Ref. [11]. If you cannot avoid the carrier, treat the samples with RNAses.
18. It is preferable to treat cfDNA samples with RNAses because DNA and RNA are sometimes extracted, and endogenous RNA may distort the size distribution profile. This was the case in one project with the Cobas® cfDNA sample preparation kit (Roche).
19. Above 150 pg/μL, the first peak of the cfDNA sample may saturate the fluorescence detector (fluorescence greater than 50 RFU; this is almost the case in Fig. 6b). It is then necessary to dilute the sample in TE 1×.
20. Thanks to the support of Agilent, we propose a macro to export all the curves of a run in a single step upon request.
21. To use other data treatment tools than BIABooster Analytics software, follow the three main steps: (i) curve smoothing (using a size window of 1–5 s), (ii) time/size and

fluorescence/concentration calibration based on the fluorescence trace of the reference ladder and linear interpolation between the peaks of the ladder, (iii) apply fluorescence/concentration calibration to the fluorescence trace of the samples, and integrate the signal between the desired time intervals. Use the time/size calibration curve to know the time interval corresponding to the desired size interval. The Agilent sequence given herein analyzes one reference ladder sample every five samples to ensure excellent analytical results. If you do not need maximum precision, for instance, if you just want to check the degradation of your sample, you can perform your analyses using a reference ladder every ten samples.

References

1. Diaz LA, Bardelli A (2014) Liquid biopsies: genotyping circulating tumor DNA. J Clin Oncol 32:579–586
2. Zonta E, Nizard P, Taly V (2015) Assessment of DNA integrity, applications for cancer research. Adv Clin Chem 70:197–246
3. Bendich A, Wilczok T, Borenfreund E (1965) Circulating DNA as a possible factor in oncogenesis. Science 148:374–376
4. Ranchon H et al (2016) DNA separation and enrichment using electro-hydrodynamic bidirectional flows in viscoelastic liquids. Lab Chip 16:1243–1253
5. Andriamanampisoa C-L, Bancaud A, Boutonnet-Rodat A et al (2018) BIABooster: online DNA concentration and size profiling with a limit of detection of 10 fg/μL and application to high-sensitivity characterization of circulating cell-free DNA. Anal Chem 90: 3766–3774
6. Chami B, Socol M, Manghi M, Bancaud A (2018) Modeling of DNA transport in viscoelastic electro-hydrodynamic flows for enhanced size separation. Soft Matter 14: 5069–5079
7. Malbec R, Chami B, Aeschbach L et al (2019) μLAS: sizing of expanded trinucleotide repeats with femtomolar sensitivity in less than 5 minutes. Sci Rep 9:23
8. Tijunelyte I, Teillet J, Bruand P et al (2022) Hybridization-based DNA biosensing with a limit of detection of 4 fM in 30 s using an electrohydrodynamic concentration module fabricated by grayscale lithography. Biomicrofluidics 16:044111
9. El Messaoudi S, Rolet F, Mouliere F, Thierry A (2013) Circulating cell free DNA: preanalytical considerations. Clin Chim Acta 424:222–230
10. Chamieh J, Martin M, Cottet H (2015) Quantitative analysis in capillary electrophoresis: transformation of raw electropherograms into continuous distributions. Anal Chem 87: 1050–1057
11. Warton K, Graham L-J, Yuwono N, Samimi G (2018) Comparison of 4 commercial kits for the extraction of circulating DNA from plasma. Cancer Genet 228–229:143–150

Chapter 3

On-Chip Magnetic Extraction of Circulating Cell-Free DNA from Biological Samples

Karla Perez-Toralla, Iago Pereiro, Sonia Garrigou, Fahima Di Federico, Monica Araya-Farias, Charlotte Proudhon, François-Clémént Bidard, Jean-Louis Viovy, Valérie Taly, and Stéphanie Descroix

Abstract

In recent years, the analysis of circulating cell-free DNA (cfDNA) containing tumor-derived DNA has emerged as a noninvasive means for cancer monitoring and personalized medicine. However, the isolation of cfDNA from peripheral blood has remained a challenge due to the low abundance and high fragmentation of these molecules. Here, we present a dynamic Magnetic ExTRactiOn (METRO) protocol using microfluidic fluidized bed technology to isolate circulating cfDNA from raw biological materials such as undiluted serum. This protocol maximizes the surface area for DNA binding within the chip in order to capture short DNA fragments. It uses only a few μL of sample and reagents. The protocol can be automated, and it is fully compatible with sensitive DNA amplification methods such as droplet-based digital PCR (ddPCR).

Key words Microfluidic fluidized bed, Circulating cell-free DNA, Circulating tumor DNA, Solid-phase extraction, Cancer

1 Introduction

The extraction of circulating cell-free DNA (cfDNA) from biological samples can be an effective alternative to invasive tissue biopsies for the detection and analysis of cancer-related biomarkers [1–3]. Circulating tumor DNA (ctDNA), which represents a small fraction of total cfDNA, is a valuable and specific biomarker for disease detection as most cancer genetic materials carry somatic mutations which can be found in ctDNA [4–6]. Microfluidic-based technologies such as targeted next-generation sequencing (NGS) and droplet-based digital PCR (ddPCR) are currently

Valérie Taly and Stéphanie Descroix equally contributed to this chapter.

Valérie Taly et al. (eds.), *Microfluidics Diagnostics: Methods and Protocols*, Methods in Molecular Biology, vol. 2804,
https://doi.org/10.1007/978-1-0716-3850-7_3,

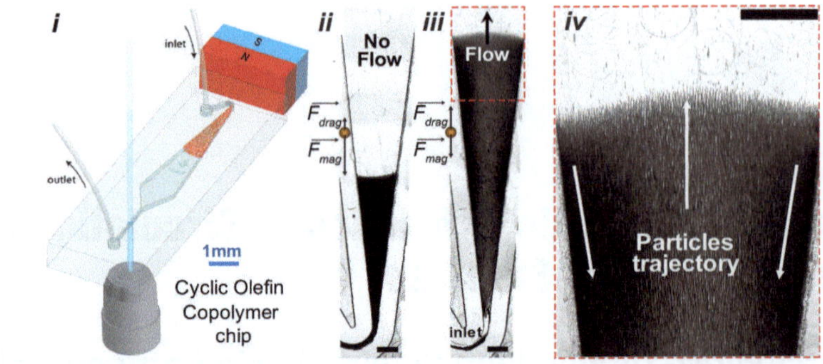

Fig. 1 Schematic of the setup for the METRO procedure using the microfluidic fluidized bed (i). Micrographs of the fluidized bed before (ii) and during (iii) sample perfusion on chip. Magnified view showing the recirculation of beads (iv). Scale bars: 500 μm. Reproduced from Perez-Toralla et al. [22] with permission from Elsevier©

being used for the highly sensitive detection and quantification of ctDNA [7–12], but their performance is still limited by pre-analytical steps performed prior to DNA analysis (e.g., sample collection and DNA extraction procedures) [13–16]. Indeed, it remains challenging to isolate efficiently ctDNA from blood as ctDNA is present at low concentrations (typically few ng/mL), and it is highly fragmented (in the 50–1000 bp size range), in particular at early stages of cancer [13, 14, 17].

Here, we report the implementation of a microfluidic Magnetic ExTRactiOn protocol (METRO) for efficient DNA isolation using fluidized bed (FB) technology [18, 19], allowing to process raw biological samples, such as undiluted serum, and deliver concentrated and inhibitor-free eluted DNA samples. In this procedure, a microfluidic fluidized bed chip was fabricated in Cyclic Olefin Copolymer (COC) (Fig. 1i). We injected functionalized magnetic beads with charge-switchable amine-terminated groups into the COC device to capture negatively charged DNA by electrostatic interaction [20, 21]. We standardized the microfluidic geometry to control the recirculation of beads inside a diamond-shaped microfluidic chamber and maximize the binding affinity to the beads' solid support. We used an external magnetic source and a flow control module to control actively the porosity of the bed of beads, resulting in a global balance between hydrodynamic drag and magnetic forces (Fig. 1ii and iii). This approach allowed the processing of important volumes of samples in reduced time while avoiding clogging artifacts and providing low backflow pressure, as well as an important surface area for analyte interaction. Samples were incubated with a lysis buffer prior to injection. Following the addition of a low pH binding buffer (pH < 6), samples were perfused through the expanded fluidized bed for DNA capture. After a washing step, DNA was eluted (pH > 8.5) and collected for DNA quantification (Figs. 1iv and 2).

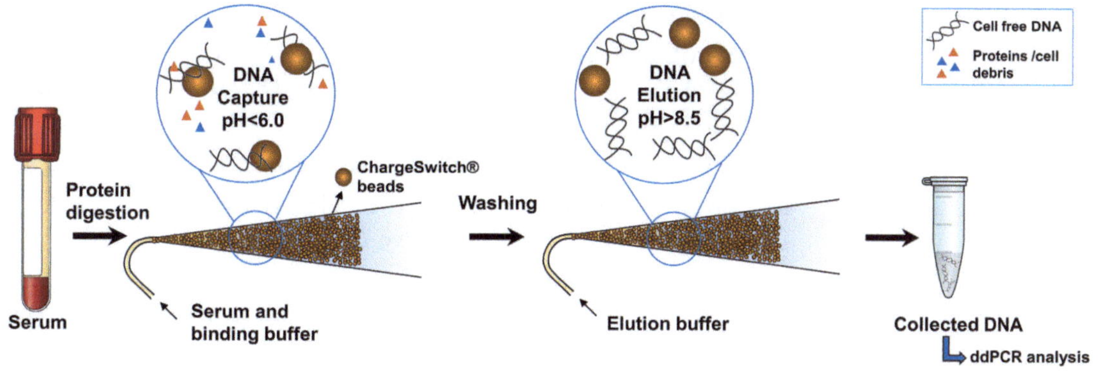

Fig. 2 Schematic showing the DNA capture and elution steps using the microfluidic fluidized bed

Implementing the METRO protocol on chip offers several advantages over the batch procedure such as the use of microliter volumes of samples and reagents. It has a high potential for automation and multiplex DNA analysis. In particular, the METRO procedure could be coupled with a second microfluidic emulsification module for the compartmentalization of the eluted DNA into picoliter droplets for the detection of rare mutational events, while reducing potential DNA loss during the transferring steps [22]. Furthermore, the METRO procedure offers a high degree of flexibility in terms of sample and elution volumes and concentration as well as in the size, number, and surface functionalities of the magnetic beads that can be implemented (e.g., to selectively capture mRNA, microRNA, methylated DNA sequences, and bacterial or viral DNA).

Here, we present the microfluidic METRO protocol for the extraction of circulating cfDNA from serum using samples spiked with genomic DNA from cancer cell lines at low concentrations. Extracted DNA can be quantified using a commercial ddPCR system. The described microfluidic-based procedure could also be applied to isolate circulating cfDNA directly from cancer patient samples.

2 Materials

Freshly prepare all solutions using analytical grade reagents and store them at room temperature (RT) (unless indicated otherwise). Filter all solutions before injecting them into the microfluidic device.

2.1 Microfluidic Device

1. Cyclic Olefin Copolymer (COC) plates (5 mm thickness) and films (254 μm thickness) (Topas® 8007, Topas Advanced Polymers).
2. Hexadecane.

3. 2-Propanol.
4. Microfluidic connections: Silicone tubing (I.D.: 0.8 mm, O. D.: 13 mm) and PTFE tubing (I.D.: 0.56 mm, O.D.: 1.07 mm).
5. Biocomponent epoxy adhesive (Araldite).
6. Solvent solution: mix 1.5 mL of Hexadecane and 8.5 mL of 2-Propanol and store in a solvent-resistant hermetic vial.
7. A master aluminum mold with the design of FB (Fig. 1) was obtained by micromilling using a 100 μm milling cutter (*see* **Note 1**).
8. Driller with a drill bit 0.5 mm in diameter.
9. Manual hydraulic press (Specac) for hot embossing.
10. Inverted microscope equipped with cooled CCD camera (*see* **Note 2**).
11. Ultrasonic bath.
12. Glass beaker.
13. Pressurized air gun.

2.2 Surface Treatment of Microfluidic Device

1. Ethanol absolute anhydrous.
2. Pluronic F127 (MW = 12,600, $EO_{99}PO_{65}EO_{99}$).
3. Pluronic solution: 1% (w/v) in water. Store at 4 °C until use.

2.3 Miniaturized Magnetic DNA Extraction on Chip

1. Phosphate-Buffered Saline (PBS) 1X.
2. ChargeSwitch gDNA Mini Tissue Kit (Life Technologies). The ChargeSwitch kit contains Proteinase K and RNAse A solutions as well as Lysis buffer (L13), Purification buffer (N5), Wash buffer (W12), and Elution buffer (E5).
3. NdFeB12 1.47 T permanent magnet (length 30 mm × height 20 mm × width 20 mm) with magnetization through the width (N50, ChenYang Technologies).
4. Binding buffer solution: mix 200 μL of PBS 1X, 140 μL Lysis buffer, and 50 μL Purification buffer.
5. Lysis mix solution: mix 60 μL of Lysis buffer, 10 μL Proteinase K solution, and 5 μL of RNAse A.
6. 1.5 mL tube magnetic separation rack.
7. Microfluidic pressure control system (MFCS, Fluigent).
8. Pressurized fluid reservoir (FLUIWELL-4C, Fluigent).
9. Bidirectional Microfluidic Flow Sensor (Flow Unit, Fluigent).
10. Vortex.
11. Centrifuge.
12. DNA LoBind Eppendorf tubes.

2.4 Biological Samples

1. Serum samples from healthy donors were centrifuged at 3000 rpm, aliquoted, and stored at −80 °C. Donor volunteers must be informed and give their written consent for the scientific use of their samples, according to procedures defined by national Health and Bio-ethics rules.
2. Cancer cell line: TP53-mutant HT-29 (c.818G > A, p.R273H) (ATCC).
3. DNA samples: Cell line DNA (from around three million of cells) extracted using QIAampDNA Mini Kit according to the manufacturer's instructions and fragmented with S220 sonicator (Covaris) to achieve DNA fragment sizes from 50 to 1100 bp. DNA samples should be aliquoted and kept at −20 °C prior to use. DNA concentration can be adjusted as needed.
4. Fluorometer (Qubit 2.0, Invitrogen, Thermo Fisher Scientific) with the dsDNA HS Assay to measure and adjust DNA concentrations as needed.

3 Methods

3.1 Microfluidic Device Design and Fabrication

1. Fabricate the aluminum master mold with the design of FB chamber (Fig. 1) (*see* **Note 3**).
2. Cut the COC plate into a 1.5 × 3 cm size and clean them with 2-Propanol. Dry the plate under pressurized air.
3. Preheat the bottom plate of the hydraulic press at 130 °C. The upper plate of the hot press should not be heated to avoid deformation of the COC.
4. Place the COC plate on top of the aluminum master with the design facing up, load into the press, and emboss the plate at 600 kPa of pressure, 120 °C for 5 min (*see* **Note 4**).
5. Cool down the resulting COC plate to 60 °C while maintaining a pressure of 600 kPa for 10 min, unload the plate from the press, and separate it from the aluminum master mold.
6. Trim and drill the embossed COC plate to create fluidic access holes. The size of the access holes should match the outer diameter of the capillary tubing.
7. Place the COC plate in a beaker containing 2-Propanol and clean it in an ultrasonic bath for 10 min. Dry under pressurized air.
8. Apply a thin film of the solvent solution on the surface of the COC plate containing the FB chamber by means of a lint-free wipe. Place a COC film on the treated surface covering the entire microfluidic design and manually apply low pressure to ensure good contact.

9. Preheated the press at 60 °C and load the chip into the press for 10 min for channel sealing.
10. Place the tubing in the inlet and the outlet of the FB chip to create fluidic connections and add epoxy adhesive around the tubing to ensure its stability (*see* **Note 5**).
11. Place the chip on the stage of microscope and visually examine the FB chip. It should be clean without clogged or deformed channels.

3.2 Surface Treatment of Chip

We performed a surface treatment using physisorbed Pluronic block copolymers to prevent nonspecific adsorption of particles or DNA molecules to the channel walls. After injection of the Pluronic F127 solution, chips were allowed to incubate for 30 minutes and rinsed by flowing PBS through the treated channels.

3.3 Serum Sample Preparation

We processed undiluted serum samples from healthy donors spiked with cancer cell line DNA (*see* **Note 6**).

1. Thaw a serum sample aliquot and centrifuge at 5000 rpm for 15 min prior to each experiment.
2. Thaw a DNA sample aliquot prior to each experiment and measure the concentration using the fluorometer.
3. Prepare the spiked serum samples by mixing the DNA sample and the serum to the required concentration (within 1 and 400 ng/mL). A serum sample with no additional cell line DNA can be used as a control for the detection of circulating wild-type genomic DNA.
4. Mix 90 μL of spiked serum with 75 μL of lysis mix solution and incubate for 1 h at RT.
5. Once the sample is digested, add 25 μL of Purification buffer to stop the lysis process, lower the pH of solution, and prepare the sample for optimal DNA binding. Then, the sample can be injected into the chip after the formation of the fluidized bed.

3.4 Bead Solution Preparation

1. Mix 2 μL of magnetic beads with 50 μL of binding buffer solution, vortex, and use the magnetic rack to pellet the beads and discard the supernatant solution.
2. Re-suspend the beads in 20 μL of binding buffer and vortex. Store at 4 °C until use.

3.5 Microfluidic Magnetic Fluidized Bed for cfDNA Extraction from Serum

We miniaturized the DNA batch extraction procedure using magnetic beads to implement a flow-through protocol on chip (METRO). We replaced all batch incubation steps of the standard protocol by steps in which samples or reagents are continuously perfused through the fluidized bed chip.

1. Connect the COC chip to the pressure control system and reservoir unit to inject all samples and reagents (*see* **Note 7**).
2. Fill the whole system at 7 μL/min with 200 μL of filtered ethanol and then rinse with 100 μL of PBS 1X.
3. Fill the whole system at 7 μL/min with 200 μL of Pluronic solution, incubate for 30 min at room temperature, and then rinse with 100 μL of PBS 1X.
4. Inject 10 μL of binding buffer at 1 μL/min.
5. Inject 10 μL of the prepared bead solution (25 μg of beads) inside the chip (*see* **Note 8**).
6. Place the NdFeB 1.47 T permanent magnet at a distance of 2 mm from the 100 μm-wide bent channels, symmetrically aligned with the axis of the main chamber (Fig. 1i).
7. Once both the beads and the magnet are in place, inject 10–20 μL of binding buffer at 1 μL/min to observe the formation of the fluidized bed of beads (*see* **Note 9**).
8. Inject 30–60 μL of prepared serum sample through the expanded fluidized bed at 0.4–0.5 μL/ min for DNA capture (*see* **Note 10**).
9. Inject 15–30 μL of Wash buffer and wash the beads inside the chip after capture at 0.5–1 μL/min to remove contaminants.
10. Inject 5–15 μL of Elution buffer at 0.2–1 μL/min and collect the eluted DNA in a DNA LoBind Eppendorf tube at the output of the chip (*see* **Note 11**).
11. Collected DNA can be measured with the fluorometer to quantify the total extracted DNA amount or by ddPCR for the high-sensitivity detection and quantification of both wild-type and mutant DNA. Samples can also be stored at −20 °C for further analysis (*see* **Note 12**).
12. Discard the COC chip containing the beads at the end of the experiment (*see* **Note 13**).

4 Notes

1. A supplementary smoothing procedure using industrial polishing cream was performed to improve the surface quality of the milled material. The mold can be obtained using other microfabrication procedures as long as it is made from metal or other thermally conductive material as required for the hot embossing process.
2. The microscope was used for optical inspection of microdevice. Visualization can be performed using different optical equipments.

3. The fluidized bed used for DNA capture was designed using a CAD software. The main chamber has a triangular shape 13 mm in length and 2.6 mm at its maximum width, with an aperture angle of 13°. The height of the chamber is 50 μm. A bent channel (100 μm in width) connects the inlet of the main chamber to the fluid reservoir. This geometry prevents beads from escaping through the inlet in no-flow conditions. The outlet is placed on the opposite side of the chamber [18].
4. Replication of the FB chamber from the aluminum master is performed by hot embossing using the manual hydraulic press. Embossing parameters should be adjusted when using different equipment or COC substrates with different grades.
5. Epoxy glue allows to keep capillary tubing in place, but other methods can also be used, e.g., frits.
6. Serum samples were spiked with clinically relevant fragment sizes (FS) and concentrations [C] of ctDNA (FS from 50 to 1100 bp and [C] = from 0 to 357 ng/mL) [23]. DNA fragment sizes were measured using the LabChip GX/GXII Microfluidic system (Perkin-Elmer) with the DNA assay 5 K reagent kit (Perkin-Elmer) and using the 2100 Bioanalyser (Agilent) with the DNA 12,000 Kit (Agilent). A variety of cancer cell lines, bearing other genetic mutations have also been used for ddPCR DNA analysis [7].
7. Samples and reagents can also be injected manually by pipetting or using syringe pumps.
8. The injection of magnetic beads was performed by disconnecting the tubing at the chamber outlet and inserting a pipette tip instead which was half-filled with binding buffer. The magnetic beads were loaded inside the tip and guided at the right location in the chamber by using (and moving) a small magnet. After beads loading, discard the tip and reconnect the outlet tubing.
9. We used 25 μg of beads (850–900 nm in diameter) to form a packed bed inside the COC microdevice. Despite the small size of the beads, we could expand the bed of beads into a fluidized regime (with homogenous bead distribution) and perfuse the different reagents using flow rates up to 1 μL/min (Fig. 1iv).
10. The volume of analyzed sample can vary according to the needs of each experiment.
11. DNA elution is driven by the recovery of the neutral charge of the bead surface at high pH, which results in a decrease of DNA binding affinity.
12. To differentiate between spiked mutant DNA (MUT) and non-mutated circulating cfDNA (wild type DNA (WT)), we used a mutation-specific assay containing allele-specific

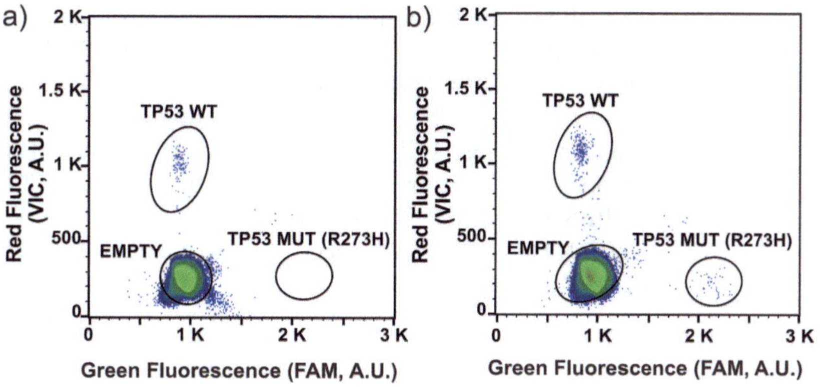

Fig. 3 Detection of R273H TP53 mutation in serum samples spiked with mutant HT29 cell line DNA (**a**, **b**). Two-dimensional histograms of the digital PCR assay for the analysis of circulating DNA extracted from healthy donors (**a**) and from spiked samples (**b**). Extracted R273H TP53 mutant DNA digitally quantified for varying spike in DNA amounts (0.0033 ng/genome copies). Custom Taqman® probes targeting wild-type (WT) sequences were conjugated to VIC (λ_{ex} 528 nm/λ_{em} 554 nm), and probes targeting mutant (MUT) sequences were labeled with 6-carboxyfluorescein (6-FAM, λ_{ex} 494 nm/λ_{em} 522 nm). Circulating cfDNA was extracted from 60 μL of serum samples using the microfluidic procedure. VIC, a proprietary dye of Life Technologies; A.U., arbitrary units. Reproduced from Perez-Toralla et al. [22] with permission from Elsevier©

hydrolysis probes (TaqMan probes targeting the R273H mutation in the TP53 gene [7]). We first evaluated the levels of background WT cfDNA (initially present in serum samples) using the METRO procedure and then the serum samples spiked with cancer cell line. We digitally quantified the number of MUT and WT-extracted genomes and compared them to the amount of DNA spiked in the different samples (Fig. 3).

13. COC devices can be discarded at the end of the experiment (as recommended when processing biological or clinical samples), or rinsed with sodium hypochlorite (10% in water) and ethanol (70% in water) solutions and dried for further use. A new surface treatment should be applied to the chip prior to each new extraction. Magnetic beads and reagents should also be replaced inside the microdevice.

Acknowledgments

The authors acknowledge the Plateforme Technologique of the Institut Universitaire d'Hématologie (IUH, Paris, France) and the Biochemistry and Molecular Biology Service (BMB168, Institut Curie, UMR168) for help with DNA analysis. This work was supported by the SIRIC CARPEM, the Ministère de l'Enseignement Supérieur et de la Recherche, the Université Paris-Descartes, the Centre National de la Recherche Scientifique (CNRS), the Institut National de la Santé et de la Recherche Médicale

(INSERM), the French Ile-de-France canceropole by ANR "Investissements d'Avenir" for Digidiag project (ANR Nanobiotechnologies, ANR-10-NANO-0002-09), the Ligue Nationale Contre le Cancer (LNCC, Program "Equipe Labelisée LIGUE"; EL2016. LNCC/VaT) as well as LABEX and EQUIPEX IPGG, and by European FP7 programs (LOVEFOOD FP7-ICT- 2011-317742, NAPES FP7-NMP-2013-604241, CATCH-U-DNA, FETOPEN H2020 and ERC Advanced Grant CellO (FP7-IDEAS-ERC-321107).

References

1. Sozzi G, Conte D, Leon M et al (2003) Quantification of free circulating DNA as a diagnostic marker in lung cancer. J Clin Oncol 21: 3902–3908. https://doi.org/10.1200/JCO.2003.02.006
2. Schwarzenbach H, Hoon DSB, Pantel K (2011) Cell-free nucleic acids as biomarkers in cancer patients. Nat Rev Cancer 11:426–437. https://doi.org/10.1038/nrc3066
3. Sonnenberg A, Marciniak JY, Rassenti L et al (2014) Rapid electrokinetic isolation of cancer-related circulating cell-free DNA directly from blood. Clin Chem 60:500–509. https://doi.org/10.1373/clinchem.2013.214874
4. Bettegowda C, Sausen M, Leary RJ et al (2014) Detection of circulating tumor DNA in early- and late-stage human malignancies. Sci Transl Med 6:224ra24. https://doi.org/10.1126/scitranslmed.3007094
5. Gorgannezhad L, Umer M, Islam MN et al (2018) Circulating tumor DNA and liquid biopsy: opportunities, challenges, and recent advances in detection technologies. Lab Chip 18:1174–1196. https://doi.org/10.1039/C8LC00100F
6. Heitzer E, Ulz P, Geigl JB (2015) Circulating tumor DNA as a liquid biopsy for cancer. Clin Chem 61:112–123. https://doi.org/10.1373/clinchem.2014.222679
7. Zonta E, Garlan F, Pécuchet N et al (2016) Multiplex detection of rare mutations by Picoliter droplet based digital PCR: sensitivity and specificity considerations. PLoS One 11: e0159094. https://doi.org/10.1371/journal.pone.0159094
8. Malapelle U, Mayo de-Las-Casas C, Rocco D et al (2017) Development of a gene panel for next-generation sequencing of clinically relevant mutations in cell-free DNA from cancer patients. Br J Cancer 116:802–810. https://doi.org/10.1038/bjc.2017.8
9. Garlan F, Laurent-Puig P, Sefrioui D et al (2017) Early evaluation of circulating tumor DNA as marker of therapeutic efficacy in metastatic colorectal cancer patients (PLACOL study). Clin Cancer Res 23:5416–5425. https://doi.org/10.1158/1078-0432.CCR-16-3155
10. Pécuchet N, Rozenholc Y, Zonta E et al (2016) Analysis of base-position error rate of next-generation sequencing to detect tumor mutations in circulating DNA. Clin Chem 62:1492–1503. https://doi.org/10.1373/clinchem.2016.258236
11. Geng Z, Li S, Zhu L et al (2020) "Sample-to-answer" detection of rare ctDNA mutation from 2 mL plasma with a fully integrated DNA extraction and digital droplet PCR microdevice for liquid biopsy. Anal Chem 92: 7240–7248. https://doi.org/10.1021/acs.analchem.0c00818
12. Ou C-Y, Vu T, Grunwald JT et al (2019) An ultrasensitive test for profiling circulating tumor DNA using integrated comprehensive droplet digital detection. Lab Chip 19:993–1005. https://doi.org/10.1039/C8LC01399C
13. Diaz LA Jr, Bardelli A (2014) Liquid biopsies: genotyping circulating tumor DNA. J Clin Oncol 32:579–586. https://doi.org/10.1200/JCO.2012.45.2011
14. El Messaoudi S, Rolet F, Mouliere F, Thierry AR (2013) Circulating cell free DNA: preanalytical considerations. Clin Chim Acta 424: 222–230. https://doi.org/10.1016/j.cca.2013.05.022
15. van Ginkel JH, van den Broek DA, van Kuik J et al (2017) Preanalytical blood sample workup for cell-free DNA analysis using droplet digital PCR for future molecular cancer diagnostics. Cancer Med 6:2297–2307. https://doi.org/10.1002/cam4.1184
16. Solar Vasconcelos JP, Boutin M, Loree JM (2022) Circulating tumor DNA in early-stage colon cancer: ready for prime time or needing refinement? Ther Adv Med Oncol 14:

17588359221143976. https://doi.org/10.1177/17588359221143975

17. Heitzer E, Haque IS, Roberts CES, Speicher MR (2019) Current and future perspectives of liquid biopsies in genomics-driven oncology. Nat Rev Genet 20:71–88. https://doi.org/10.1038/s41576-018-0071-5
18. Pereiro I, Bendali A, Tabnaoui S et al (2017) A new microfluidic approach for the one-step capture, amplification and label-free quantification of bacteria from raw samples. Chem Sci 8: 1329–1336. https://doi.org/10.1039/C6SC03880H
19. Trapiella-Alfonso L, Alexandre L, Fraichard C et al (2019) VEGF (vascular endothelial growth factor) functionalized magnetic beads in a microfluidic device to improve the Angiogenic balance in preeclampsia. Hypertension 74:145–153. https://doi.org/10.1161/HYPERTENSIONAHA.118.12380
20. Hagan KA, Meier WL, Ferrance JP, Landers JP (2009) Chitosan-coated silica as a solid phase for RNA purification in a microfluidic device. Anal Chem 81:5249–5256. https://doi.org/10.1021/ac900820z
21. Tiwari AP, Satvekar RK, Rohiwal SS et al (2015) Magneto-separation of genomic deoxyribose nucleic acid using pH responsive Fe3O4@silica@chitosan nanoparticles in biological samples. RSC Adv 5:8463–8470. https://doi.org/10.1039/C4RA15806G
22. Perez-Toralla K, Pereiro I, Garrigou S et al (2019) Microfluidic extraction and digital quantification of circulating cell-free DNA from serum. Sensors Actuators B Chem 286: 533–539. https://doi.org/10.1016/j.snb.2019.01.159
23. Mouliere F, Robert B, Peyrotte EA et al (2011) High fragmentation characterizes tumour-derived circulating DNA. PLoS One 6: e23418. https://doi.org/10.1371/journal.pone.0023418

Chapter 4

Isolation of Extracellular Vesicles from Cell Culture and Blood Through Nano-Targeted DLD Microfluidic Device

Marie Gaillard, Nicolas Sarrut-Rio, Catherine Pudda, François Boizot, Yoann Roupioz, and Aurélie Thuaire

Abstract

Extracellular vesicles (EVs) are secreted by cells and found in biological fluids such as blood, with concentration correlated with oncogenic signals, making them attractive biomarkers for liquid biopsy. The current gold-standard method for EVs isolation requires an ultracentrifugation (UC) step among others. The cost and complexity of this technique are forbiddingly high for many researchers, as well as for routine use in biological laboratories and hospitals. This chapter reports on a simple microfluidic method for EVs isolation, based on a microfluidic size sorting technique named Deterministic Lateral Displacement (DLD). With the design of micrometric DLD array, we demonstrated the potential of our DLD devices for the isolation of nano-biological objects such as EVs, with main population size distribution consistent with UC technique.

Key words Extracellular vesicles, Isolation, Deterministic lateral displacement, Size sorting, Microfluidic

1 Introduction

Extracellular vesicles (EVs) are lipid nanoparticles secreted by cells for communication purposes that circulate in biological fluids such as blood [1]. They convey information from a parental cell to another cell thanks to proteins displayed on their surface and their DNA/RNA content. As their abundance and the expression of specific surface proteins have been shown to vary in correlation with specific pathologies, mainly cancers, they are considered as promising biomarkers in the field of liquid biopsies [2]. EVs isolation is however a technological challenge. It is currently mostly addressed using differential centrifugation, as described by Théry et al. [3]. However, this method involves the use of an ultracentrifuge apparatus, which turns out to be labor intensive, time and cost-consuming, and requires significant sample volumes to ensure

Valérie Taly et al. (eds.), *Microfluidics Diagnostics: Methods and Protocols*, Methods in Molecular Biology, vol. 2804,
https://doi.org/10.1007/978-1-0716-3850-7_4,

sufficient reproducibility. Because of the significant number of limitations due to differential centrifugation, the International Society of Extracellular Vesicles (ISEV) has encouraged the development of new methods for EVs isolation in 2018 [4], fostering samples and measurement protocols standardization, sufficient reproducibility, speed, and simplicity for operators.

Microfluidic-based techniques may address such issues by enabling the analysis of a low biological sample volume and in a simpler manner than centrifugation-based approaches. Deterministic Lateral Displacement (DLD) is a size-sorting microfluidic technique based on a slightly disoriented pillar array [5]. DLD network geometry defines a critical diameter (D_c) according to which particles of larger sizes are deviated according to the pillar array orientation while smaller ones are not. During the last decade, this technique has been successfully applied to the sorting of micrometric biological objects: blood cells [6], circulating tumor cells [7], and bacteria [8], among others. As recently reported by Hochstetter et al., nowadays the main challenge of DLD is the sorting of nanoparticles [9]. Previous works have already demonstrated the ability to sort nanovesicles using nanoscale DLD arrays but required restrictive preparation steps [10, 11]. Practically, EVs are isolated and concentrated by ultracentrifugation before being sorted by nanoscale DLD. However, there is still a lack of simple operation of DLD onto EV samples. The approach we describe in this chapter aims at sorting nanometric particles using micrometer-sized pillar array (Fig. 1), rather than nanometer-sized structures which are way more challenging to fabricate than our initial micrometric DLD sorting system [12]. Micrometric pillar arrays enable to decrease both the fluidic resistance and the overall processing time, and in the meantime significantly increase the flow rate through the device by comparison to nanometric structures. This DLD approach allows our system to be generic for a large range of nanometric biological objects. The proposed method is based on

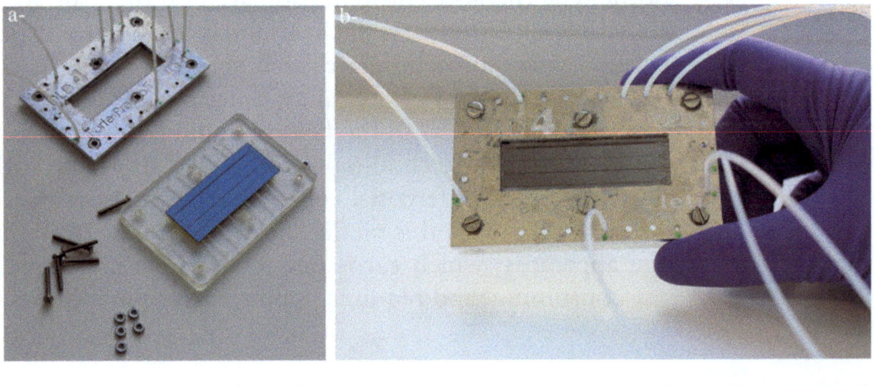

Fig. 1 Pictures of nano-targeted DLD device: (**a**) different components of the DLD device, (**b**) assembled DLD device

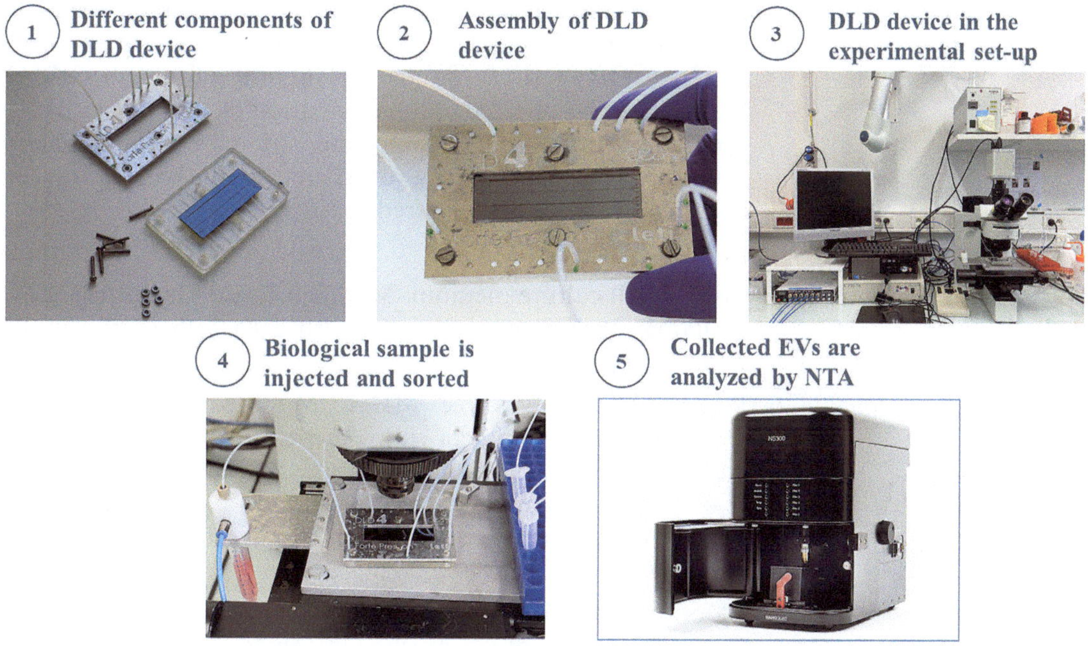

Fig. 2 Scheme of EVs sorting steps with DLD

the association of (i) two low-centrifugation steps, meant to guarantee the absence of large particles/cell debris that could generate clogging in the DLD channel, with (ii) the sorting processing through DLD system. In this chapter, we propose a detailed practical procedure to successfully carry out EVs sorting (Fig. 2), either from culture media or blood, with the DLD technique. It will cover the description of preliminary steps, such as biological sample preparation and DLD device fabrication, as well as the fluidic implementation of EVs sorting through DLD and characterization of the sorted samples through Nano-Tracking Analysis (NTA). A comparison with UC classical isolation protocol will also be provided, showing similar isolation of EVs subpopulations with a comparable precision, but processing lower sample volumes, which is an asset for biological samples.

2 Materials

2.1 Biological Samples and Reagents

1. THP-1 cells from ATCC.
2. THP-1 cell culture medium: Rosewell Park Memorial Institute Medium (RPMI) 1640 Glutamax media supplemented with 10% Foetal Bovine Serum FBS (v:v) and 1% PenStrep (10,000 U/mL and 10 mg/mL Penicillin and Streptomycin, respectively) (v:v).

3. EVs-free RPMI 1640: RPMI 1640 Glutamax media supplemented with 20% Foetal Bovine Serum FBS (v:v) and 1% PenStrep (v:v) and centrifuged at 100,000 g for 18 h at 4 °C. Supernatant is collected, diluted (dilution factor = 2) in RPMI 1640 Glutamax media supplemented with 1% PenStrep (10,000 U/mL and 10 mg/mL Penicillin and Streptomycin, respectively) (v:v) and is stored at −20 °C.
4. MCF-7 cells from ATCC.
5. MCF-7 cell culture medium: Modified Eagle Medium (MEM) Glutamax media supplemented with 10% FBS (v:v) and 1% PenStrep (v:v).
6. Preparation of EVs-free MEM: Modified Eagle Medium (MEM) Glutamax media supplemented with 10% FBS (v:v) and 1% PenStrep (v:v) and centrifuged at 100,000 g for 18 h at 4 °C. Supernatant is collected, diluted (dilution factor = 2) in MEM Glutamax media supplemented with 1% PenStrep (v: v), and stored at −20 °C.
7. Blood samples: 5 × 10 mL of blood (heparin as anticoagulant). Blood donor volunteers must be informed and give their written consent for the scientific use of their blood, according to procedures defined by national Health and Bio-ethics rules.
8. PBS filtered onto 0.22 μm.

2.2 Nano-Targeted DLD Microfluidic Device

2.2.1 DLD Silicon Device

The DLD critical diameter D_c has to be adjusted according to the size range of the population of interest. For EVs from THP-1 and MCF-7 cells as well as for EVs from plasma, the size range is 50–300 nm. In this chapter, we will then focus on a DLD channel with a D_c of 350 nm, keeping in mind that it should be adjusted depending on the size of the targeted population.

Reaching a submicrometric D_c, while targeting deep channels (i.e., high pillar height to diameter aspect ratio) to optimize the sample flow rate, requires the use of silicon technologies for the fabrication of DLD channels. In addition to the reproducibility and reliability brought by these fabrication techniques, it is possible to address micrometric and submicrometric scales and high aspect ratio structures. The main features we are using are described in Table 1 and the main fabrication steps are illustrated in Fig. 3.

A DLD chip is composed of (i) a silicon substrate including the DLD fluidic channel and the fluidic inlets/outlets and (ii) a glass cover closing the fluidic system. These 200 mm substrates are assembled at wafer scale in a cleanroom with an anodic bonding process. Preliminary to this step, the DLD pillar network is patterned in the silicon substrate through a DRIE process (Deep Reactive Ion Etching), leading to pillars aspect Ratio of 15. After anodic bonding, the fabrication process is completed with the patterning of fluidic inlets/outlets (DRIE process) in the silicon substrate. 200 mm wafers are then diced into chips.

Table 1
Main features of DLD silicon chip $D_c = 350$ nm

DLD $D_c = 350$ nm	
Pillar diameter (D_P)	2.5 μm
Interpillar gap (G)	2.5 μm
Period (N)	70
Channel depth (H)	37.5 μm
Array length (L)	44 mm
Array width (W)	575 μm

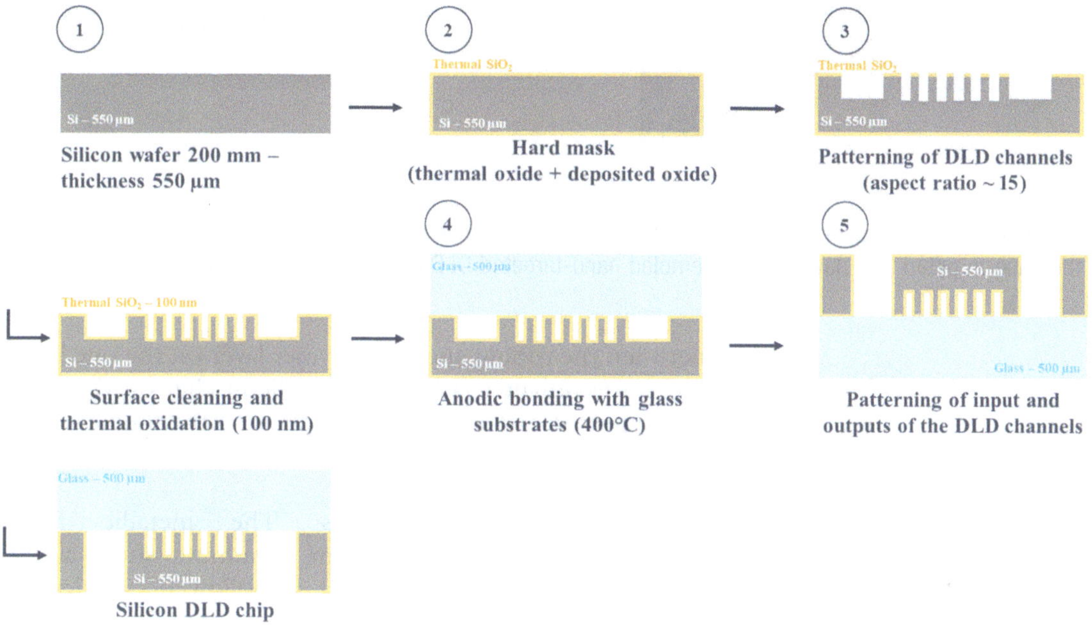

Fig. 3 Global DLD chip process flow

These chips can however neither be directly connected to a fluidic sample nor an actuation system (syringe pumps, pressure controllers), and they require a fluidic packaging as an interface, referred to as a "microfluidic cartridge."

2.2.2 Microfluidic Cartridge in Cyclic Olefin Copolymer (COC)

This microfluidic cartridge consists in a polymer substrate on which the silicon chip is laid and a screwed metallic frame, featuring holes for tubing insertion. Between these two parts, ferrules and home-made silicon gaskets ensure fluidic sealing (Fig. 4).

The polymer substrate is made from two thermally sealed COC polymer layers and houses micromachined fluidic channels (200 μm in diameter). The screwed metallic frame is meant to ensure a good

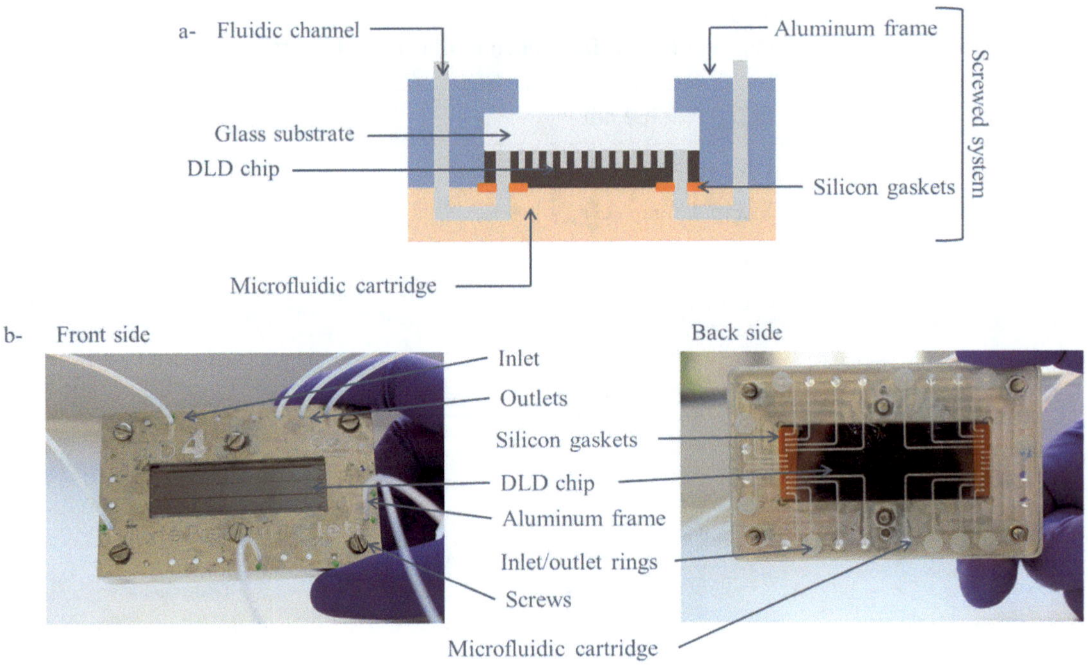

Fig. 4 Fluidic packaging of nano-targeted DLD microfluidic device: (**a**) schematic of the cross-sectional view of the nano-targeted DLD device, (**b**) assembled nano-targeted DLD device on the front and back side

sealing between the tubing and the polymer layer on the one hand, and between the polymer layer and the silicon chip on the other hand, while a pressure of a few bars is applied on the fluidic sample at the inlet.

2.2.3 Device Assembly

1. Aluminum frame and screws: The metallic frame (8.5 mm*5.4 mm) is made from aluminum sheet and is in-house manufactured. Its center is hollowed (window of 5.5 mm*1.8 mm) for optical monitoring of the DLD channel during fluidic filling. It includes holes for screw, as well as for tubing connection with the microfluidic cartridge. Sealing between tubing and microfluidic cartridge is ensured by ferrules.
2. Tubing: The fluidic tubing used is PTFE 1/16 inch × 0.8 mm. About 15 cm length is used for each inlet/outlet.
3. Inlet and outlet ferrules: The ferrules used for inlet and outlet rings are compatible with 1/16 inch outside diameter tubing.
4. Silicon gaskets: Silicon gaskets are molded using silicon Nusil MED-6010, with a mold corresponding to the fluidic inlets/outlets of the silicon chip (holes of 500 μm in diameter).
5. Greiner® 15 mL centrifuge tubes, used as containers for sample injection into the microfluidic system inlet.

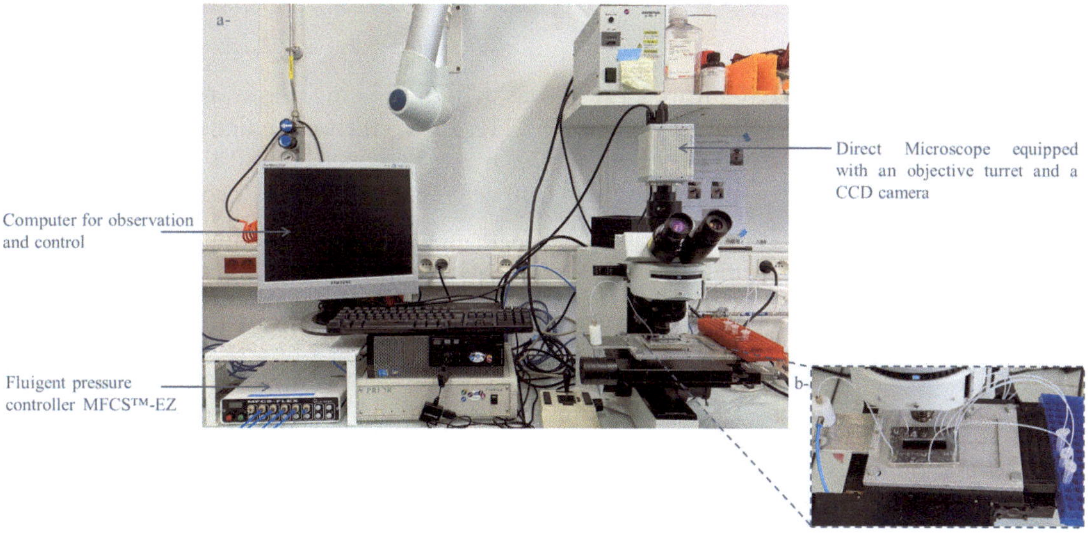

Fig. 5 DLD microfluidic device in the experimental set-up

6. Adapted cap for centrifuge tube: The centrifuge tube is closed with a Teflon cap, enabling both a pneumatic interface (for the pressurization of the sample) and a fluidic interface (for the sample injection toward DLD). This cap is homemade and considers the pneumatic and fluidic connectors diameters for the holes sizing (Fig. 5).
7. 2 mL Eppendorf® tubes, used as containers for the collection of the sorted sample at the microfluidic system outlets.

2.3 Equipment

1. Low-speed centrifuge (up to 2500 g) for 50 mL Falcon® tubes.
2. High-speed centrifuge (Beckmann, Optima L-100 K).
3. Direct Microscope equipped with an objective turret (2.5× and 10×) and a sensitive CCD camera (Hamamatsu Orca Flash 4.0LT+) controlled by computer through software (HCImageLive 4.4.2).
4. Fluigent pressure controller MFCS™-EZ controlled by computer through software Maesflo 3.4.0.
5. Nanosight NS300, Malvern Panalytical controlled by computer through software NTA 3.4 and associated syringe pump (Nanosight syringe pump, Malvern Panalytical), controlled by computer through software NTA 3.4.

2.4 Consumables

1. Luer filter Amicon® Ultra-15 K 100 K.
2. Luer filter 1.2 μm.
3. Filter 0.22 μm.
4. 50 mL Falcon® tubes
5. T75 (75 cm^2) flasks.

3 Methods

3.1 Preparation of Biological Samples

3.1.1 THP-1 Cell Culture

1. Culture THP-1 cells in T75 (75 cm^2) flasks (two flasks per run, 2 × 12 mL) using THP-1 cell culture medium. Change culture media every 48–72 h and pass cells at 80% of confluence in maximum (130,000 cells/cm^2 or 800,000 cells/mL).
2. 48 h before EVs isolation, remove culture medium by pipetting and add EVs-free RPMI 1640 + 10% FBS + 1% PenStrep fresh medium.
3. Collect the content of two T75 flasks of THP-1 cells (2 × 12 mL) and centrifuge the supernatant at 300 g for 10 min at room temperature in 15 mL centrifuge tubes.
4. Pellet the cells, collect the resulting supernatant, transfer it into two 15 mL centrifuge tubes, and centrifuge at 2000 g for 10 min at room temperature. Dead cells and large cell debris are pelleted: collect the supernatant (2 × 10 mL) and concentrate it onto Amicon® Ultra-15 K 100 K filter (concentration factor = 5) and filter onto a luer connected 1.2 μm filter on a 10 mL syringe. Approximate collected volume is 2 mL: place it into a Greiner® 15 mL tube.
5. The biological fluid, placed into Greiner® tube, is ready to be injected (*see* Subheading 3.2). Biological samples are preferably used up to 6 h after preparation at room temperature.

3.1.2 MCF-7 Cell Culture

1. Culture MCF-7 cells in T75 flasks (two flasks per run, 2 × 12 mL) using MCF-7 cell culture medium. Culture media are changed every 48–72 h and cells are passed at 80% of confluence at maximum (13,000 cells/cm^2 or 80,000 cells/mL).
2. 48 h before EVs isolation, remove culture medium by pipetting and add EVs-free MEM + 10% FBS +1% PenStrep fresh medium.
3. Collect the content of two T75 flasks of MCF-7 cells and centrifuge the supernatant at 300 g for 10 min at room temperature in two 15 mL centrifuge tubes.
4. Pellet the cells, collect the resulting supernatant, transfer it into two 15 mL centrifuge tubes, and centrifuge at 2000 g for 10 min at room temperature. Dead cells and large cell debris are pelleted: collect the supernatant (2 × 10 mL) and concentrate it onto Amicon® Ultra-15 K 100 K filter (concentration factor = 5) and filter onto a luer connected 1.2 μm filter on a 10 mL syringe. Approximate collected volume is 2 mL: place it into a Greiner® 15 mL tube.
5. The biological fluid, placed into Greiner® tube, is ready to be injected (*see* Subheading 3.2). Biological samples are preferably used up to 6 h after preparation at room temperature.

3.1.3 Preparing Human Plasma from Blood

1. Pool together 50 mL of blood samples in the same 50 mL centrifuge tube. Centrifuge the pooled sample for 15 min at 2500 g at room temperature to remove blood cells (red blood cells, white blood cells, and platelets).
2. Collect the supernatant and centrifuge a second time, for 15 min at 2500 g at room temperature to remove residual blood cells.
3. Slowly collect the supernatant (plasma) by pipetting and avoiding the collection of any precipitated material (about 40–50% of the total volume).
4. Once the plasma is collected, dilute at 1:10 in PBS.
5. Filter the biological sample onto a luer connected 1.2 μm filter on a 10 mL syringe and place it into a Greiner® 15 mL tube (EVs are concentrated at about 10^{10}/mL).
6. The biological fluid, placed into the Greiner® tube, is ready to be injected, see Method (Subheading 3.2). Biological samples are preferably used up to 6 h after preparation at room temperature.

3.2 Implementation of DLD Microfluidic Device and EVs Sorting (Fig. 2)

1. Take a DLD silicon chip from storage box.
2. Precisely align silicon gaskets with the silicon chip inlets and outlets (*see* **Note 1**).
3. Equip the metallic frame with appropriate inlet and outlet tubing associated with ferrules (Fig. 2, picture 1).
4. DLD chip is sandwiched between the metallic frame and the plastic microfluidic cartridge, and the system is manually screwed (*see* **Note 2**) (Fig. 2, picture 2).
5. Set the system under the microscope associated to the sensitive camera (Fig. 2, picture 3).
6. The previously prepared biological sample (see Subheadings 3.1.1–3.1.3) is added to a specific Greiner® 15 mL centrifuge tube and connected to the DLD system with the inlet tubing, which is adjusted through the centrifuge tube via the adapted cap to ensure the sealing (Fig. 2, picture 4).
7. Connect the DLD outlets to Eppendorf® tubes (through the outlet tubings) for sample collection after sorting. Considering a DLD channel with a critical diameter of 350 nm, EVs are expected to flow without being deviated and to be collected in the non-deviated outlets. Larger particles are expected to be separated from EVs while being deviated by the pillar array up to the deviated outlet.
8. Inject the biological sample into the DLD device by increasing the pressure up to 2 bars with the pressure controller, for about 3 min (filling step) (*see* **Notes 3** and **4**).

9. Monitor the filling progress into the channel using the microscope and camera.
10. When the DLD system is properly filled with the sample, raise injection pressure up to 4 bars, the sorting step begins. It lasts approximately 6 h for a 2 mL sample (flowrate in the order of 350 μL/h).
11. Sorted samples are collected in Eppendorf® tubes at DLD outlets.

3.3 Size Characterization of Extracellular Vesicle Samples Using NTA

EVs collected in non-deviated outlets are characterized in size by NTA (*see* **Note 5**).

3.3.1 Recommendations Before Analysis

An ideal volume of 1 mL and an ideal concentration of about 50 particles/image (acceptable if higher than 10–15 particles/image) and corresponding to a sample concentration of 10^7 to 10^9 particles/mL are usually recommended for a relevant analysis. Isolated EVs may thus have to be diluted in filtered PBS. UC-isolated EVs are typically diluted with a factor of 1000 before NTA analysis while DLD-isolated EVs are usually not diluted.

3.3.2 Analysis

1. Set the NTA-associated syringe pump to have a flow rate of 50 units.
2. Adjust the focus and camera level according to particle type (camera level is usually between 14 and 16 for EVs).
3. The number of particles per frame is in the order of 50 otherwise the sample dilution has to be adjusted.
4. Three captures of 60–90 s are usually performed.
5. Data are then processed, and the detection threshold is chosen. This detection threshold corresponds to the number of pixels to be considered as a particle (usually set at 7–8 in our study).

At the end of this protocol, EVs size profile allows EVs sample characterization and informs about the efficiency of EVs isolation through DLD device. EVs subpopulations can be compared for different biological samples or for different isolation protocols. In our case, we can note that the NTA size profile of EVs isolated by DLD device extends from 50 to 300 nm (Fig. 6) and that the major EVs subpopulation around 100 nm is very similar for both DLD and UC approaches. However, lower sample volumes can be processed using the DLD microfluidic system, which is an asset for biological samples. Samples volumes and concentrations used in both approaches are summarized in Table 2.

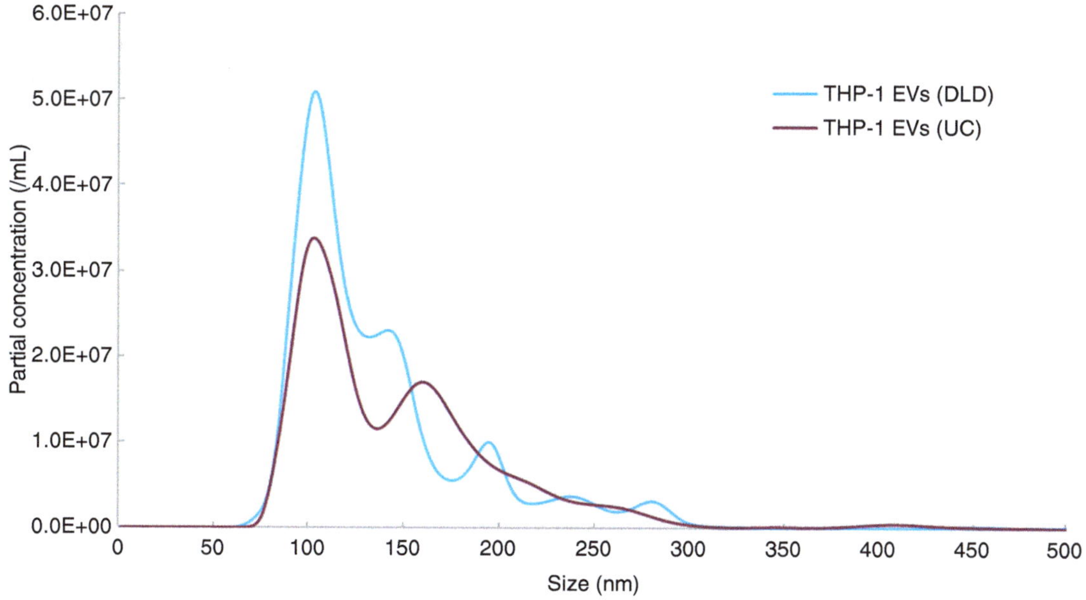

Fig. 6 Comparison of size and concentration profiles for EVs from THP-1 cells isolated either by DLD ($D_c = 350$ nm) (blue curve) or by the standard UC method (purple curve)

Table 2
Summary of cell media needed and collected EVs by UC and DLD isolation methods

	Starting cells				Collected EVs			
	Biological sample preparation		Injected volume (mL)		UC		DLD	
	Concentration (cells/mL)	Volume (mL)	UC	DLD	Concentration (EVs/mL)	Volume (μL)	Concentration (EVs/mL)	Volume (mL)
THP-1 cell line	800,000	24	24	2	7×10^{10}	500	3×10^{9}	1.5
MCF-7 cell line	80,000	24	24	2	2×10^{11}	500	6×10^{9}	1.5
Blood	5,000,000	50	25	2	2×10^{11}	500	7×10^{9}	1.5

4 Notes

1. Sealing tips with silicon gaskets: As mentioned above, silicon gaskets are fabricated at CEA-Leti using silicon Nusil MED-6010. If these types of gaskets are not available, O-ring gaskets can be adapted to each inlets/outlets of the DLD silicon chip.

2. Tips for the screwing of DLD device: A fine tightening of the screws of the microfluidic device is necessary to ensure a good seal while avoiding breaking the DLD silicon component which is relatively fragile. It is recommended to screw the screws crosswise and to check the good crushing of the gaskets.
3. Avoiding bubbles in DLD: The tubbing is gently placed in the biological sample and the injection pressure is fixed at 2 bars to avoid the appearance of bubbles and to ensure the good filling of the microfluidic channel.
4. Avoiding clogging in DLD: In microfluidics, clogging is a crucial issue. In DLD and particularly in nano-targeted DLD applications, clogging must be avoided for the longest time possible. We propose to prevent it by following the recommendations established in this paper (i.e., concentration of particles in biological samples of about 10^8 part/mL and filtration of the buffer). We evaluated that preconditioning is not required, and the sample is directly injected into the DLD device.
5. Ensuring a good NTA characterization: NTA analysis is optimal when the number of particles per image is about 50 so that each particle can scatter the light relative to itself independently. Since EVs are biological particles, their refractive index is relatively low implying a camera level setting between 14 and 16.

Acknowledgments

The work is part of a PhD thesis (M. G.) funded by the French Laboratory of Excellence ANR program Labex ARCANE (ANR-17-EURE-0003). This project was also partly supported by the FET Proactive H2020 Viruscan project (731868). Authors would like to thank particularly Christine Chatellard from IBS for her guidance with the NTA and her help for the easy IBS access.

References

1. EL Andaloussi S, Mäger I, Breakefield XO et al (2013) Extracellular vesicles: biology and emerging therapeutic opportunities. Nat Rev Drug Discov 12:347–357
2. Eguchi A, Kostallari E, Feldstein AE et al (2019) Extracellular vesicles, the liquid biopsy of the future. J Hepatol 70:1292–1294
3. Théry C, Amigorena S, Raposo G et al (2006) Isolation and characterization of exosomes from cell culture supernatants and biological fluids. Curr Protoc Cell Biol 30:3–22
4. Théry C, Witwer KW, Aikawa E et al (2018) Minimal information for studies of extracellular vesicles 2018 (MISEV2018): a position statement of the International Society for Extracellular Vesicles and update of the MISEV2014 guidelines. J Extracell Vesicles 7:1535750
5. Inglis DW, Morton KJ, Davis JA et al (2008) Microfluidic device for label-free measurement of platelet activation. Lab Chip 8:925
6. Civin CI, Ward T, Skelley AM et al (2016) Automated leukocyte processing by

microfluidic deterministic lateral displacement. Cytometry A 89:1073–1083

7. Au SH, Edd J, Stoddard AE et al (2017) Microfluidic isolation of circulating tumor cell clusters by size and asymmetry. Sci Rep 7:2433
8. Holm SH, Beech JP, Barrett MP et al (2016) Simplifying microfluidic separation devices towards field-detection of blood parasites. Anal Chem 8:3291–3300
9. Hochstetter A, Vernekar R, Austin RH et al (2020) Deterministic lateral displacement: challenges and perspectives. ACS Nano 14: 10784–10795
10. Smith JT, Wunsch BH, Dogra N et al (2018) Integrated nanoscale deterministic lateral displacement arrays for separation of extracellular vesicles from clinically-relevant volumes of biological samples. Lab Chip 18:3913–3925
11. Wunsch BH, Hsieh KY, Kim S-C et al (2021) Advancements in throughput, lifetime, purification, and workflow for integrated nanoscale deterministic lateral displacement. Adv Mater Technol 6:2001083
12. Pariset E, Parent C, Fouillet Y et al (2018) Separation of biological particles in a modular platform of cascaded deterministic lateral displacement modules. Sci Rep 8:17762

Chapter 5

Microchip for Immunomagnetic Sorting of Circulating Tumor Cells (CTCs)

Lucie Descamps, Emmanuelle Laurenceau, Sophie Cavassila, Léa Payen, Damien Le Roy, and Anne-Laure Deman

Abstract

Circulating tumor cells (CTCs) isolated directly from whole blood opens new perspectives for cancer monitoring and the development of personalized treatments. However, due to their rarity among the multitude of blood cells, it remains a challenge to recover them alive with high level of purity, i.e., with few remaining white blood cells, and in a time frame compatible with the clinical context. Microfluidic chips have emerged as promising tools to address these challenges. We propose a two-step workflow including a pre-enrichment step, performed by a size-based pre-enrichment system, and a purification step, performed by an immunomagnetic chip. Here, we describe the protocol for the fabrication of the immunomagnetic microchip, the preparation of the sample, and the procedure for injection into the microchip allowing the sorting of the CTCs.

Key words Circulating Tumor Cells (CTC), CTC recovery, Sample purification, Immunomagnetic sorting, Microfluidic chip, Composite polymer, Microfabrication

1 Introduction

Circulating tumor cells (CTCs) have received significant attention over years for their potential clinical significance. The isolation of CTCs directly from blood, as a liquid biopsy, could be used for early cancer diagnosis, patient outcome prognosis, and treatment efficacy monitoring, paving the way for personalized cancer medicine [1–3]. Microfluidic chips have emerged as promising tools to isolate CTCs since they offer precise cell manipulation, versatile functions, and cost-effective fabrication [4–6]. Microfluidic-based isolation technologies rely on the differences between CTCs and blood cells, either in physical properties (size, deformability, density, or electric charges) [7, 8] or biological properties (surface marker expression) [9, 10]. Despite the recent advance in the conception of microchips, physicians, and biologists do not use

Valérie Taly et al. (eds.), *Microfluidics Diagnostics: Methods and Protocols*, Methods in Molecular Biology, vol. 2804,
https://doi.org/10.1007/978-1-0716-3850-7_5,

these microchips on a routine basis to sort CTCs because of the challenges that still need to be overcome in a clinical context. The comparable size of CTCs to some white blood cells, and the heterogeneities of tumor markers explain the difficulty of achieving high purity and performing subsequent analysis [11]. Moreover, the initial volume of processed whole blood as well as the separation time must be compatible with clinical patient management. Finally, the system must provide easy handling for the end-users to ensure its integration into routine clinical studies. We opted for the promising strategy of combining two stages of CTC sorting, one based on physical properties and one based on biological properties to take advantage of the benefits of both [12]. We have therefore developed a two-step CTC sorting protocol consisting of a pre-enrichment step, which is performed by a size-based pre-enrichment system (ClearCell FX1®) [13], and a purification step, which is achieved through a microchip. For this second step, we propose a microfluidic device integrating dense arrays of permanent micro-magnets for negative enrichment of CTCs by trapping magnetically labeled white blood cells (WBCs) and isolating viable CTCs for downstream characterization. The micro-magnets are obtained through an original fabrication approach, based on magnetic polymer composites, that breaks with standard microfabrication techniques [14–16]. We magnetically structured a mixture of NdFeB microparticles in a polydimethylsiloxane (PDMS) matrix, to obtain an array of high aspect ratio micro-structures that generate magnetic forces up to several nanoNewtons in the close vicinity of their surface [17]. These micro-magnets, whose density is about 1500 traps/mm^2, all act as individual traps for labeled WBC, while they have no influence on the CTC. We used cancer cell lines as mimicking CTC models (mCTCs) and demonstrated their successful recovery while preserving their viability and integrity for subsequent analyses. From a 300 μL sample containing 3.10^5 WBCs and 2.10^4 mCTCs, the microchip achieved depletion of all magnetically labeled WBCs (85% of WBCs were successfully labeled) and ensured a CTC recovery rate of 81% at a flow rate of 2 mL/h. In addition, we showed its compatibility with conventional biological studies, including 2D and 3D cell culture, as well as phenotypic and genotypic analyses. Finally, the two-step separation workflow that combines a size-based pre-enrichment system (ClearCell FX1®) and a subsequent purification step with the microchip led to high throughput (7.5 mL blood sample processing in less than 4 h), high purity (947 WBCs per mL remaining, 99.99% depletion rate) and a CTC recovery rate of 71% [18].

2 Materials

2.1 Composite Preparation

1. Polydimethylsiloxane (PDMS), Sylard™ 184, bi-component: a base and a curing agent.
2. NdFeB microparticles from Magnequench (0.5–7 μm) of irregularly shaped flakes.
3. Microstructuring system: one NdFeB magnet (Supermagnete), $60 \times 30 \times 15$ mm^3 in size generating a magnetic induction of 0.3 T in the composite membrane.
4. Magnetization system: two NdFeB magnets (Supermagnete), $50.8 \times 50.8 \times 25.4$ mm^3 in size and spaced 2 mm apart employing a homemade holder and generating a magnetic induction of 1.2 T. The generated magnetic field can be enhanced thanks to the addition of a U-shaped soft steel part to guide the flux between the external poles of the magnets on both sides.
5. Kapton adhesive tape.
6. Glass slides.
7. Ethanol.
8. Mortar.
9. Precision balance.
10. Scraper.
11. Cutter.
12. Oven
13. Laminar flow unit.

2.2 Device Assembly

1. Dry Photoresist, 100 μm-thick (Etertec).
2. Developer solution: 10 g/L sodium carbonate in distilled water.
3. Trichloro(1H,1H,2H,2H-perfluorooctyl)silane.
4. Biopsy puncher (1.25 mm diameter).
5. Insolation lamp.
6. Photomask.
7. Orbital shaker.
8. Desiccator.
9. Oxygen plasma cleaner.

2.3 Sample Preparation

1. Pluronic solution: 1% Pluronic F-108 in PBS
2. Filtered phosphate buffer saline (PBS).

3. Labeling solution: 2 mM EDTA (Ethylene diamine tetra acetic acid (EDTA)) in 2% BSA-PBS (Bovine serum albumin—BSA) (*see* **Note 1**).
4. Anti-CD45 and anti-CD15 antibodies-conjugated superparamagnetic nanoparticles (MasterBeads Carboxylic Acid 0215, Ademtech). The nanoparticles (NP) used are 500 nm in diameter and composed of a magnetic core (approximately 70% iron oxide) encapsulated by a hydrophilic polymer shell with carboxyl groups on its surface. Covalent attachment of human-anti-CD45-DL650 and human anti-CD15-AlexaFluo647 fluorescent antibodies (R&D Systems) to the nanoparticles' surface was performed following the manufacturer's protocol (*see* **Note 2**).
5. 48-well plates (CytoOne).
6. Incubator with shaking plate (MS-100 Thermoshaker Incubator).
7. Centrifuge.

2.4 Microfluidic Chip Manipulation

1. Pressure controller (FLOW EZ™, Fluigent) (*see* **Note 3**).
2. Fluigent connectors (P-CAP 2 mL kit) to ensure airtightness of the Eppendorf tube.
3. Tubing with an external diameter of 1/16″ OD, i.e., 1.58 μm, and an internal diameter (ID) of 500 μm (*see* **Note 4**).
4. 1.5 mL Eppendorf safe-lock tube.
5. Two NdFeB magnets (Supermagnete), $25 \times 8 \times 2\ mm^3$ in size, generating a magnetic induction of 0.2 T in the channel (1 mm above their surface), housed in a homemade plastic holder so that they are positioned identically for each microsystem (spaced 4 mm apart and placed 2 cm from the chamber entrance).

3 Methods

3.1 Fabrication of the Microfluidic Chip

The system consists of two PDMS parts, which are then bonded together to form a sealed system. The lower part, the composite membrane, integrates the magnetic micro-traps. The upper PDMS part integrates the microfluidic channels and the trapping chamber, as well as the holes allowing the connection of tubes for the injection of samples into the chip. The fabrication of the systems should preferably be done under a laminar flow unit.

1. Prepare PDMS by mixing the base and the curing agent at a 10: 1 ratio (*see* **Note 5**).

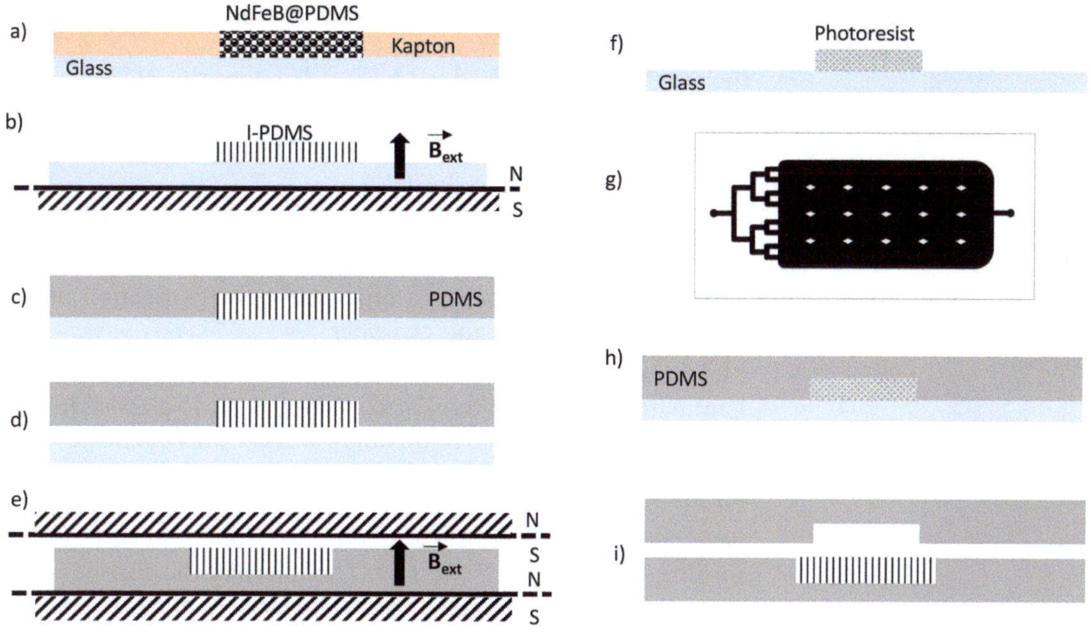

Fig. 1 (**a–i**) Flowchart of the fabrication steps of the microchip

2. Add 2 wt% of the NdFeB micro-particles in prepared PDMS and thoroughly mix, for a few minutes, in a mortar, to obtain the NdFeB@PDMS composite.
3. Cut in 100-μm-thick Kapton adhesive tape (*see* **Note 6**) a rectangular area whose surface corresponds to that of the trapping area, 40×20 mm^2 in our case. (*see* **Note 7**). The density of magnetic traps being of the order of 1500 traps/mm^2, the total number of traps in the chip is about 10^6.
4. Clean a glass slide by soaking it in ethanol and drying it with dry air. It will serve as a support for the fabrication of the composite membrane.
5. Stuck the Kapton film on the center of the cleaned glass slide.
6. Pour the NdFeB@PDMS composite on the mold and align the thickness of the composite with the thickness of the Kapton film using a scraper (Fig. 1a).
7. Gently remove the Kapton film, leaving a 100-μm-thick composite layer, and put it in an oven to cure it at 70 °C for 2 h in a magnetic induction of 0.3 T (Fig. 1b) supplied by the micro-structuring system (*see* **Note 8**). During curing, NdFeB particles are free to move in the liquid polymer and self-organize driven by magnetic dipolar interactions. They form chains of NdFeB particles oriented along the magnetic field lines. After curing, NdFeB particles are immobilized in the polymer matrix.

8. Pour pure PDMS mix on the composite membrane, which is placed in a petri dish, to increase its thickness up to 2 mm. Cure the membrane at 70 °C for 2 h (Fig. 1c). Cut the PDMS with a cutter, making sure that the membrane has the dimensions of the glass slide, then peel it off (Fig. 1d).
9. Magnetize composite membrane in the magnetizing system (Fig. 1e). (*see* **Note 9**). The composite membrane is completed.
10. Clean another glass slide by soaking it in ethanol and drying it with dry air. It will serve as a support for the fabrication of the microfluidic channels and chamber.
11. Make the mold of the microfluidic device by soft-lithography. Laminate the dry photoresist onto the cleaned glass slide and insolate (14 mW for 8 s) through a mask on which the fluid channels and the trapping chamber are printed. Then, cure at 70 °C for 2 min the channel patterns and reveal them in a developer bath, previously heated to 60°, for 8 min with orbital shaking (Fig. 1f) (*see* **Note 10**). The microfluidic channel consists of a trapping chamber with a surface equal to that of the magnetic composite (40 × 20 mm^2). Micro-channels that feed the chamber are distributed over its width to limit the formation of bubbles during the filling of the channels. The microfluidic channel mold design is reported in Fig. 1g (*see* **Note 11**).
12. Silanize the mold using silane in a desiccator for at least 3 h at room temperature to prevent undesired bonding of PDMS to the mold.
13. Pour pure PDMS mix on the mold and cure at 70 °C for 2 h (Fig. 1h). Cut the PDMS with a cutter to the dimensions of a glass slide and remove it from the mold.
14. In order to keep the tubes well maintained in the microchip, the thickness of the microchip inlet and outlet is increased by 1 mm, by irreversibly bonding a reticulated PDMS sheet (5 × 7 × 1 mm^3) using O_2 plasma surface treatment.
15. After curing, punch the interconnection holes using the biopsy puncher. The top PDMS microfluidic part is completed.
16. Bond the microfluidic channel with the composite membrane irreversibly using O_2 plasma surface treatment (Fig. 1i). Be careful to align the trapping chamber on the composite area during bonding (*see* **Note 12**). Place the assembled device in the oven for 5 min at 70 °C.

3.2 WBCs Labeling

1. Recover the sample from the pre-enrichment step. In our case, an average of 3.10^5 WBC remains. Spin down and resuspend the pellet in a final volume of 300 μL of the labeling solution (*see* **Note 13**).

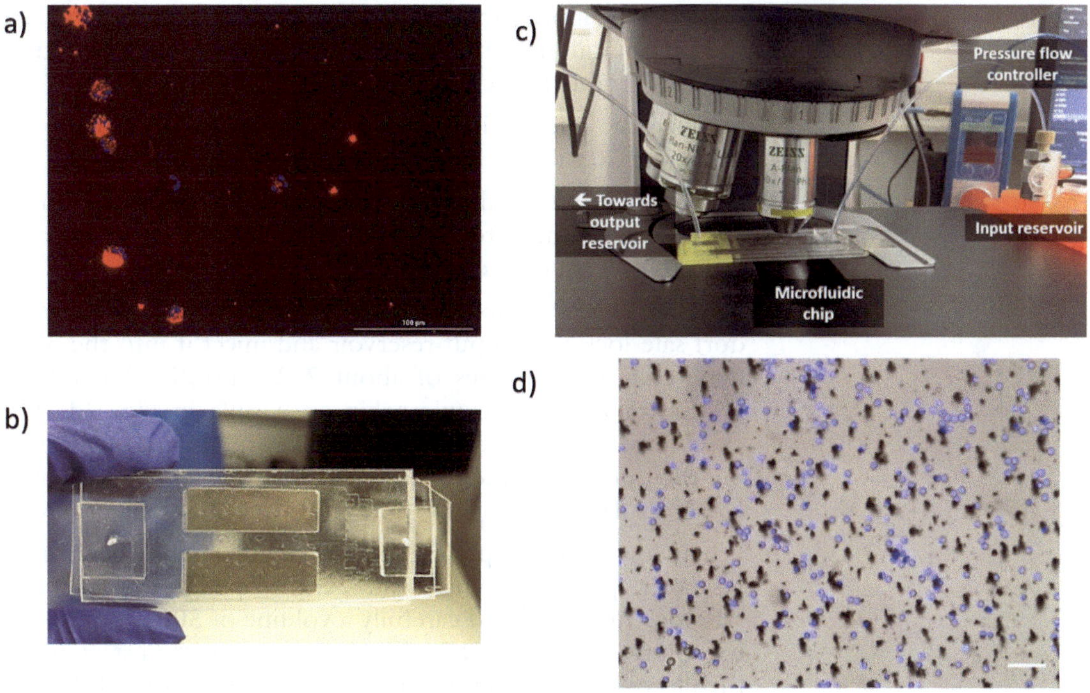

Fig. 2 (**a**) Fluorescence microscope image of labeled WBCs (nucleus stained in blue) whose surface is studded with magnetic nanoparticles (in red). Different levels of labeling can be observed depending on surface marker expression. Excessed nanoparticles can also be found. (**b**) Picture of the fabricated microchip, with the plastic support that holds the two permanent milli-magnets ($25 \times 8 \times 2$ mm^3 each) below the trapping chamber. (**c**) Picture of the experimental setup. The microfluidic system is placed under the microscope. The input reservoir is connected to the pressure controller for pressure-driven injection and to the channel input through FEP tubing. The processed sample is collected in a dedicated output reservoir. (**d**) Microscopic images of captured WBCs on the magnetic micro-traps which are integrated into the microchip (only the fluorescence of the nuclei in blue is shown). (Adapted with permission from Descamps et al. [18], http:/creativecommons.org/licenses/by-nc-nd/3.0/)

2. WBCs are labeled with antibody-conjugated magnetic nanoparticles at a concentration of 400 NP/WBC and 100 NP/WBC for anti-CD45 and anti-CD15 antibodies, respectively. WBCs and functionalized nanoparticles are incubated in a 48-well plate at 37 °C for 30 min under a gentle vortex agitation of 200 rpm. (*see* **Note 14**). Over the hundreds of observed WBCs, 85% of them were labeled (Fig. 2a).

3.3 CTC Purification in the Microchip

The dimensions of the chip are compatible with an optical microscope to monitor the injection and trapping steps if needed.

1. Position the trapping chamber on top of the two external millimeter-sized permanent magnet (Fig. 2b) (*see* **Note 15**). These milli-magnets attract the WBCs circulating at the top of the channel, toward the micro-traps.

2. Connect the input tubing to the pressure controller and secure the output tubing with tape into a clean 1.5 mL Eppendorf tube for sample collection (Fig. 2c).
3. Sterilize the microfluidic device by injecting 70% ethanol for 10 min at 20 mbar.
4. Before cell injection, flush the device with Pluronic solution for 20 min at 10 mbar to remove ethanol and coat the channel walls to prevent cell adhesion.
5. Transfer the cell sample (300 μL volume) in a 1.5 mL Eppendorf safe-lock tube input reservoir and inject it into the magnetic chip at flow rates of about 2–2.5 mL/h (2–6 mbar). WBCs trapped in the microchip are visible in Fig. 2d. The exact flow rate is determined after each experiment by measuring the output volume (collected within the experiment time) (*see* **Note 16**).
6. When less than 10 μL of the sample remains in the input reservoir, pause the injection, gently open the Eppendorf safe-lock tube, and add carefully a volume of 300 μL of Pluronic solution (*see* **Note 17**). The injection can then start again, the Pluronic will push the remaining cells in tubing and channel toward the output reservoir.
7. Recovered CTCs in the Eppendorf are available for subsequent analysis (counting, Immunofluorescence staining, 2D or 3D culture, FISH...). The whole protocol depletes 99.99% of WBCs, leaving approximately 947 WBCs per mL of initial blood.

4 Notes

1. Adding EDTA and BSA is critical during white blood cell labeling to avoid cellular agglomeration. The mixture can be adjusted with a base (NaOH) to obtain an EDTA BSA-PBS solution with a pH between 7 and 8.
2. Functionalized nanoparticles can be kept for 1 week in the fridge.
3. Other systems are possible to inject the cells, such as syringe pumps. Care must be taken to limit the sedimentation of cells in syringes or tubing since they may not be injected.
4. This tubing's internal diameter was chosen for cell experiments to avoid clogging since cancer cells can form clusters with sizes reaching 350 μm.
5. PDMS begins to crosslink once the mixing is done, so mix it with the particles without delay and pour the composite into the mold.

6. Other films than Kapton can be used. The only constraint is to ensure that the PDMS cross-links when in contact with this film.
7. The trapping area can be resized according to the wanted number of traps and knowing the obtained micro-traps density with the composite.
8. It is important to position the trapping area in the center of the magnet so that the magnetic field lines seen by the magnetic micro-particles are perpendicular to the surface.
9. The magnetic field can be generated by other means, like electromagnets, providing a uniform, intense enough field to approach the saturation of the magnetically hard particles, and large enough to magnetize the entire trapping area.
10. The mold of the fluidic system can be realized with different techniques, by UV lithography or laser lithography, with a liquid or dry resin of 100 μm thickness, or in plastic by 3D printing.
11. Support pillars of diamond shape are integrated into the trapping chamber to avoid its collapse due to the large ratio between the chamber width (20 mm) and height (100 μm).
12. This step is critical because it is not possible to reposition the two PDMS slices once they have been in contact after the O_2 plasma surface activation.
13. To calculate the amount of NP, we considered an average concentration of 3.10^5 WBCs in the pre-enrichment outlet.
14. For experiments requiring fluorescent imaging, WBC nuclei were marked with Hoechst by adding one drop of the marker solution to the solution of WBCs and NPs within the well, before the incubation.
15. The plastic support for magnets also allows to maintain the microchip, which can bend because of the flexibility of PDMS.
16. Since the injection is controlled in pressure, the corresponding pressure (in mbar) for the desired flow rate (e.g. in mL/h) was first estimated by calculation, then experimentally assessed by measuring the collected liquid volume after a certain injection time.
17. Tubing and chamber volumes being estimated at 180 μL, therefore 300 μL of 1% Pluronic was injected.

Acknowledgments

The authors acknowledge funding for this work from LUTON, PACK AMBITION RECHERCHE AuRA and from PURECHIP,

Cancer-PCSI, ITMO Cancer (Aviesan, Inserm). The authors acknowledge J. Degouttes for the fabrication of the devoted plastic holders and all the support staff of the Nanolyon technology platform. All the staff of the CIRCAN team from the Hospices Civils de Lyon is also gratefully acknowledged for their help on biological sample preparation and AztraZeneca for their CIRCAN's manpower support.

References

1. Lin E, Cao T, Nagrath S, King MR (2018) Circulating tumor cells: diagnostic and therapeutic applications. Annu Rev Biomed Eng 20: 329–352. https://doi.org/10.1146/annurev-bioeng-062117-120947
2. Quandt D, Zucht HD, Amann A et al (2017) Implementing liquid biopsies into clinical decision making for cancer immunotherapy. Oncotarget 8:48507–48520
3. Chemi F, Mohan S, Guevara T et al (2021) Early dissemination of circulating tumor cells: biological and clinical insights, vol 11. Front, Oncol, p 11
4. Liang W, Liu J, Yang X et al (2020) Microfluidic-based cancer cell separation using active and passive mechanisms. Microfluid Nanofluidics 24:26. https://doi.org/10.1007/s10404-020-2331-x
5. Yu X, Wu N, Chen F et al (2019) Engineering microfluidic chip for circulating tumor cells: from enrichment, release to single cell analysis. Trends Anal Chem 117:27–38. https://doi.org/10.1016/j.trac.2019.03.027
6. Descamps L, Le Roy D, Deman AL (2022) Microfluidic-based technologies for CTC isolation: a review of 10 years of intense efforts towards liquid biopsy. Int J Mol Sci 23:1981. https://doi.org/10.3390/ijms23041981
7. Kalyan S, Torabi C, Khoo H et al (2021) Inertial microfluidics enabling clinical research. Micromachines 12:1–43. https://doi.org/10.3390/mi12030257
8. Sollier E, Go DE, Che J et al (2014) Size-selective collection of circulating tumor cells using vortex technology. Lab Chip 14:63–77. https://doi.org/10.1039/c3lc50689d
9. Poudineh M, Aldridge PM, Ahmed S et al (2017) Tracking the dynamics of circulating tumour cell phenotypes using nanoparticle-mediated magnetic ranking. Nat Nanotechnol 12:274–281. https://doi.org/10.1038/nnano.2016.239
10. Xu CM, Tang M, Feng J et al (2020) A liquid biopsy- guided drug release system for cancer theranostics: integrating rapid circulating tumor cell detection and precision tumor therapy. Lab Chip 20:1418–1425. https://doi.org/10.1039/d0lc00149j
11. Alix-Panabières C, Pantel K (2014) Challenges in circulating tumour cell research. Nat Rev Cancer 14:623–631
12. Jack R, Hussain K, Rodrigues D et al (2017) Microfluidic continuum sorting of subpopulations of tumor cells via surface antibody expression levels. Lab Chip 17:1349–1358. https://doi.org/10.1039/C6LC01496H
13. Warkiani ME, Khoo BL, Wu L et al (2016) Ultra-fast, label-free isolation of circulating tumor cells from blood using spiral microfluidics. Nat Protoc 11:134–148. https://doi.org/10.1038/nprot.2016.003
14. Descamps L, Le Roy D, Tomba C, Deman AL (2021) Magnetic polymers for magnetophoretic separation in microfluidic devices. Magnetochemistry 7:100
15. Bae YM, Jeong B, Kim J et al (2019) Array of 3D permanent micromagnet for immunomagnetic separation. J Micromech Microeng 29: 085007. https://doi.org/10.1088/1361-6439/ab259f
16. Royet D, Hériveaux Y, Marchalot J et al (2017) Using injection molding and reversible bonding for easy fabrication of magnetic cell trapping and sorting devices. J Magn Magn Mater 427:306–313. https://doi.org/10.1016/j.jmmm.2016.10.102
17. Descamps L, Audry MC, Howard J et al (2021) Self-assembled permanent micromagnets in a polymer-based microfluidic device for magnetic cell sorting. Cell 10:1734
18. Descamps L, Garcia J, Barthelemy D et al (2022) MagPure chip: an immunomagnetic-based microfluidic device for high purification of circulating tumor cells from liquid biopsies. Lab Chip 22:4151. https://doi.org/10.1039/D2LC00443G

Part III

Microchips for Protein Markers Analysis/Detection

Chapter 6

Enhancing the Efficiency of Conventional Surface Immunoassays Within Standard Labware Using Microscale Flows

Iago Pereiro, Anna Fomitcheva-Khartchenko, and Govind V. Kaigala

Abstract

In this chapter, we present the design and fabrication of a device and implementation of a protocol to realize increased efficiency of immunoassays within microtiter plates. The device, WellProbe, is a 3D-structured probe that can be used to deliver precise flows at the bottom of standard well plates to establish concentric areas of shear stress intensities using hydrodynamically confined flows. The protocols involve both operation and data analysis.

Key words Advection, Mass transport, Microtiter plate, Multiplexing, Open-space microfluidics, Surface assays

1 Introduction

Assays performed on solid phases are prevalent in chemistry and the life sciences, including immunoassays (e.g., enzyme-linked immunosorbent assay, and ELISA), molecular patterning, and cell assays [1–3]. Most such assays are performed in standard labware, particularly in well plates. Well plates ensure compartmentalization, disposability, reproducibility, and compatibility with laboratory equipment including microscopes, readers, and pipettors. However, their geometry is a barrier to efficient mass transport, which generally relies on diffusion. Because biomolecules present low diffusion coefficients, diffusion-driven transport often leads to suboptimal reaction kinetics, as the analyte gets depleted in the layer adjacent to the reaction site [4].

Creating flows over the reaction area improves mass transport in assays and significantly reduces time-to-result and/or enhances sensitivity, sometimes by several orders of magnitude [5]. Orbital shakers are the conventional solution to create moderate flows in well plates, but they generate nonuniform flows mostly limited to

Valérie Taly et al. (eds.), *Microfluidics Diagnostics: Methods and Protocols*, Methods in Molecular Biology, vol. 2804,
https://doi.org/10.1007/978-1-0716-3850-7_6,

the liquid bulk far from the reaction surface, resulting in modest signal increases of low reproducibility [2]. Alternative microfluidic solutions to enhance mass transport in surface reactions have so far relied on customized devices, incompatible with commercially available well plates [4, 6].

WellProbe is a 3D-structured probe that can deliver precise flows at the bottom of standard well plates, establishing several concentric areas of defined shear stress intensities [7]. Performing well plate-based surface assays with WellProbe provides the following advantages: (i) it reduces assay time and increases sensitivity, (ii) it expands the linear quantitation range of the assay with several local kinetic conditions, (iii) it can identify saturated conditions, artifacts, and false positives/negatives, and (iv) it enables rapid liquid switching for multistep protocols, making intermediate washing steps unnecessary. Additionally, it allows high-level multiplexing in single-well tests by coupling it with arrays of radially distributed spots.

In order to execute this type of assay, we here describe methods around device fabrication, operation, and data analysis.

2 Materials

Freshly prepare all solutions using analytical grade reagents and store them at room temperature (RT) (unless indicated otherwise). Filter all solutions before injecting into the microfluidic device.

2.1 Probe and Setup

1. WellProbe fabricated with aluminum by micromilling (PICOMAX 60 CNC, Fehlmann).
2. Shrinking tube (Distrelec).
3. Hot air dryer (200 °C).
4. PTFE tubing (OD 1/32 inch).
5. 3D-printed probe holder.
6. X, Y, and Z motorized linear stages (Zaber).
7. Microfluidic pressure control system (MFCS, Fluigent).
8. Pressurized fluid reservoir (FLUIWELL-4C, Fluigent).
9. Bidirectional Microfluidic Flow Sensor (Flow Unit, Fluigent).
10. Inverted microscope.
11. Small petri dish.
12. Ethanol absolute.
13. Distilled water.
14. Phosphate-buffered saline (PBS) 1X.

2.2 Immunoassays

1. Inkjet spotter (GeSiM).
2. Desiccator.
3. Microtiter plate with transparent bottom.
4. Bovine serum albumin.
5. Capture antibody and fluorescent detection antibody of choice. Store at 4 °C until use.

3 Methods

3.1 Device and Setup Design Considerations and Fabrication

WellProbe injects reagents of interest through an inner inlet channel onto the surface of a well in a microtiter plate. The reagent flows radially away from this inlet and is aspirated by a surrounding groove, defining a circular area of reagent confinement where chemical and biological reactions can take place (Fig. 1a).

A probe is a solid cylinder of diameter smaller than the well diameter to address, containing at least an inner central channel for flow delivery. This central channel is connected to a fluid reservoir through a tubing, with a flowrate provided by a syringe pump or a pressure source. Alternatively, a manual or robot pipettor can be used to directly inject liquid into the channel. The injected liquid is distributed radially from the center of the well.

Below we describe design considerations of the probe and setup.

3.1.1 Profile Shape

The bottom surface of the probe defines a gap with the bottom of the well plate. The geometry of this gap, together with the injected flow rate, determines the flow intensity and the shear stress created at the well bottom:

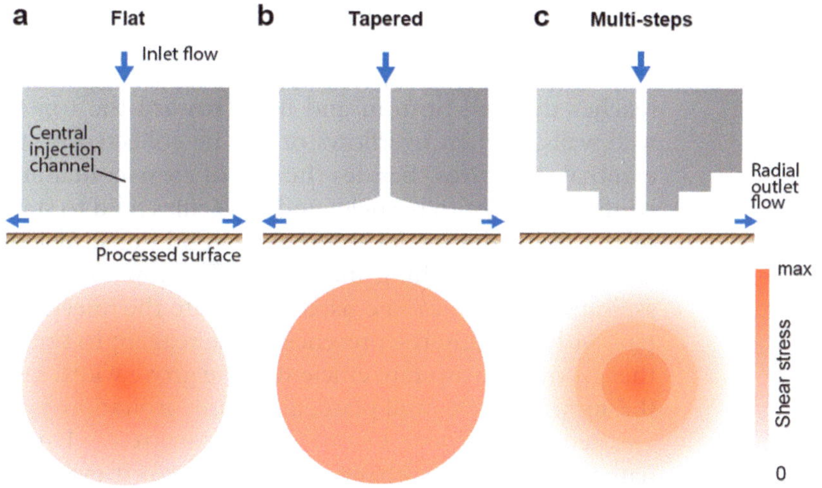

Fig. 1 Variants of WellProbe design and resulting shear stress on processed area: (**a**) flat, (**b**) tapered, and (**c**) multi-steps. The probe geometry is radially symmetrical; a central cross-section is shown

- Flat surface, the simplest geometry (Fig. 1a). The created shear stress on the well is nonuniform, with higher values near the inlet and minimal at the probe edge. This is due to the decaying flow velocity in the radial direction. For molecular assays, this can result in higher kinetics (and signal) near the center.
- Tapered surface, with a maximum gap at the inlet and minimal on the sides (Fig. 1b). With the right geometry, this can result in a uniform shear stress on the surface [8], with the gap H defined as a function of the radius from the probe center, by the equation:

$$H(r) = H_i \sqrt{\frac{r_i}{r}} \quad (1)$$

where H_i and r_i represent the height and radius of the central aperture.
- Mesas, defining concentric gaps of increasing height from the center (Fig. 1c). These can be used to create areas of similar shear stress and thus kinetic conditions, to multiplex the number of kinetic conditions per area.

3.1.2 Feet

The distance between probe and well bottom can be set with protrusions at the probe edge, their height defining the height of the gap. Such protrusions, or feet, can be distributed symmetrically, e.g., at three or four equally distanced edges. If the well bottom is deformable, this further contributes to setting the surface parallel to the probe surface. These feet should be laterally small (e.g., a few hundred micrometers) to minimally perturb the flow. Alternatively, the gap can be defined with a holder that attaches the probe at the inlet side and sits on the upper rim of the well. However, it is important to consider the fabrication tolerance of the well plate, often in the order of 100μm.

3.1.3 Outer Aspiration

In the simplest probe, the liquid injected from the probe center reaches the well bottom and flows toward the top following the well walls. It then overflows or can be collected or aspirated with external structures. Besides the risk of cross-contamination, in this configuration both particles and molecules tend to stay at the edges of the wells, due to local low flow conditions, making sequential chemistry and washing inefficient. As shown in Fig. 2, an outer aspirating groove can be used to aspirate the injected liquid simultaneously, limiting the presence of the reagent to the inner circle of this groove. To aspirate efficiently, the grove width must be at least 0.5 mm. For a 96-well plate, a uniform aspiration by the groove is ensured if the groove height is approximately >1 mm when evacuated from a single channel, or >0.5 mm if evacuated from two opposite channels.

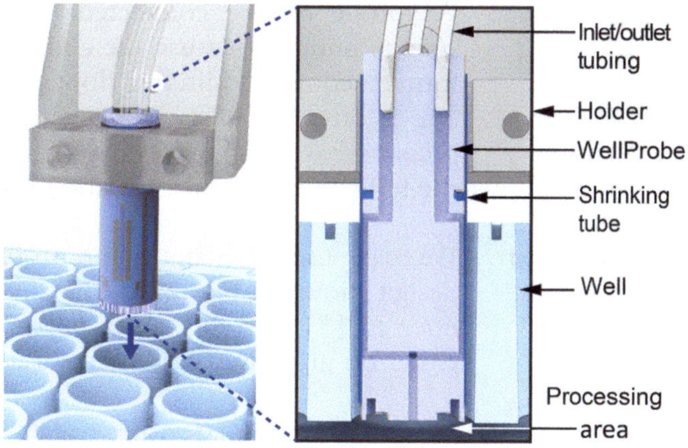

Fig. 2 WellProbe in holder before and after insertion in a well plate

3.1.4 Bubble Trap

The geometry of the probe can be optimized to reduce the probability of bubbles disrupting the experiment. In the configuration with mesas, bubbles injected through the central channel will tend to migrate toward the outer mesas due to drag and buoyancy forces, eventually escaping the probe. In the presence of a surrounding groove (as in Fig. 2), the bubble can simply sit on the ceiling of the groove, generally not perturbing the flow. The presence of two aspiration channels ensures that the flow is maintained even if one aperture is blocked by a bubble.

3.1.5 Materials and Fabrication

The probe is amenable to a range of material, including metal, polymers, and glass. Complex internal fluidics can be fabricated with 3D printing, in order to e.g., bring several liquids to the inlet, or to aspirate from the groove through multiple channels. If micromachining metal or more standard polymers, a solution to create multiple internal channels is to groove them on the probe sides and close these grooves with an outer polymer ring to create closed channels (Fig. 2). This can be done with thermally activated plastic shrinking tubes.

3.1.6 Stage

The position of WellProbe is defined with a motorized stage. In its simplest form, a single linear stage can be mounted vertically on a stable base plate. A 3D printed or micromachined holder similar to the one shown in Fig. 2 can be used as the interface between stage and probe. The design with mesas from Fig. 1c is particularly robust to deviations from a vertical alignment and can be used with this simple setup. For flat and tapered geometries, we recommend additionally adding goniometer stages to finely adjust the vertical position. Finally, to fully automate the dispensing in the different wells of a microtiter plate, two additional linear motorized stages can be added to either move the probe or the plate laterally, to position the probe in the desired well.

3.1.7 Fluidics

Wellprobe can be connected to both pressurized reservoirs or syringe pumps. Because the control of the flowrate is crucial, the former requires additional flow sensors. Since the aspiration flowrate is often significantly stronger than the injection flowrate, a pressure-based aspiration may be advantageous to avoid reaching the full volume of an aspirating syringe.

3.2 Device Operation

We make use of an aluminum probe with lateral grooves with the design shown in Fig. 2. For details and technical drawings, please refer to [7]. We focus our protocols on surface immunoassays, either on uniformly patterned surfaces or on protein microarrays.

3.2.1 Station Setup

1. The aluminum device is sealed using a polymer shrinking tube. For that, a fragment of a shrinking tube of a diameter slightly larger than the probe is cut in a length ~ 10% smaller than the device. Then by holding the device and the shrinking tube with tweezers heat is applied using a hot air dryer (200 °C) for around 30 s. Cool down the device, e.g., by immersing it in water. ATTENTION! The shrinking tube has to cover all the open lateral channels of the device.
2. Insert WellProbe vertically in the supporting 3D-printed holder. Hold the device by tightly screwing the two parts of the holder. WellProbe must be held on the top part, leaving enough device length for a complete insertion into a microtiter well plate.
3. Place WellProbe over an inverted microscope objective for better control of the reaction. Room light or light projected from the side usually helps in better visualizing the 3D structure of the probe.
4. Connect one extreme of 1/32″ tubing to WellProbe's inlet channels: one of the channels is an aspiration, while the other channels are used for injection of chemicals. Tubing can be held with Araldite glue for better connection. Connect the other end of the tubing to a disposable pressurized reservoir. The pressure can be applied by a pressure flow controller or a syringe pump. For a better control of the flow, it is recommended to insert a flowrate sensor between WellProbe and the reservoir.
5. Place a small dish under WellProbe and perform head cleaning and bubble removal by injecting ethanol during 2 min at maximum pressure (1.5 bar) through all the channels, followed by 2 min of distilled water and finally 2 min with PBS. This step removes possible dirt and reduces the probability of bubbles [9].

6. Fill one injection reservoir with the analyte and the other with PBS. Confirm that the downstream aspiration reservoir has enough space for aspiration.
7. Pre-fill all the channels with the solutions of interest. Keep the aspiration channel in distilled water to avoid bubble formation. Check the correct functioning of all channels before removing the dish and proceeding to the next step. ATTENTION: before the start of the experiment, the apex of the head has to be wet at all times. Failure to do so may result in an increased amount of bubbles because of drying.

3.2.2 Perform a Whole-Surface Immunoassay (Monoplex)

Prior to initiating the experiment:

1. Functionalize microtiter plate wells with primary antibody for 1 h. Antibody concentration has to be adjusted based on the antibody used; a recommended starting point is 50μg/mL.
2. Remove the excess of antibodies by washing with PBS three times.
3. Block the wells using 1% BSA in PBS for 30 min.
4. Remove excess blocking solution by washing the wells 3 times with PBS.

Protocol with WellProbe:

5. Place the microtiter plate on a fluorescent microscope and position WellProbe over the plate.
6. With a syringe, fill the well of interest with 300μL of PBS. It is better if some liquid protrudes out of the well.
7. Carefully approach WellProbe using the Z-stage to the well until the liquid of both the well and WellProbe touch. Continue submerging WellProbe until it reaches the well bottom, which can be observed by a subtle displacement of the WellProbe to one side. Excess buffer will leak in the space between the wells.
8. Confirm the presence of buffer surrounding WellProbe inside the well. Begin to aspirate at −100μL/min. ATTENTION: control the presence of PBS in the well and add more before it runs out. Low buffer levels will result in the presence of bubbles and might compromise the experiment in the whole well.
9. Reduce aspiration to 41μL/min and start the injection of fluorescently labeled antibody at 40μL/min (ratio 1:1) and leave for 1–6 min. The fluorescence of the antibody will indicate the limits of the confinement. ATTENTION: instabilities during this step of the experiment may indicate the presence of bubbles, particles; or issues with the flow rates. Changes in the response time of the flow after switching parameters may

indicate issues with the experiment, as it should be almost immediate.

10. Stop the fluorescently labeled antibody injection and initiate injection of PBS at 40μL/min for 1 min.
11. Stop the injection of PBS and reduce the aspiration rate to 20μL/min. Slowly pull the device out of the well. When WellProbe is almost out of the well, stop the aspiration completely and then proceed to remove the device.
12. Proceed to wash the head to avoid cross-contamination between wells. Flush the device with water at 1.5 bar for 30 s, then with ethanol for 1 min and then with water for 1 min. ATTENTION: a more aggressive wash might be required in certain cases. In this case, replace the ethanol wash with bleach. (*see* **Note 1**).

3.2.3 Perform an Immunoassay with Microarrays (Multiplex)

1. Create a circular pattern on the software of an inkjet spotter. By using a microdrop dispenser tip, deposit 10–15 drops of primary antibody (50μg/mL) on each spot. ATTENTION: test the program on a glass slide before doing it on a microtiter plate to avoid breaking the inkjet spotter tip.
2. Dry the spots for 1 h in a desiccator.
3. Proceed as in Subheading 3.2.2 starting at **step 3** (*see* **Note 2**).

3.2.4 Hydrodynamic Reconfigurability to Expand the Quantitation Range

Under certain circumstances, samples with a high concentration of analytes can lead to signal saturation. To increase further the upper range of measurable analyte concentrations, it is possible to use the dynamic reconfigurability of the flow confinement to address an additional external zone of the surface (zone 4 in Fig. 3) for a different incubation time. For this, the flow confinement can be shrunk by increasing the aspiration while the injection is kept constant. We recommend a ratio 1:3 (e.g., using 40μL/min for injection and − 120μL/min for aspiration) for the flow rates. Incubation times need to be experimentally adjusted, but an

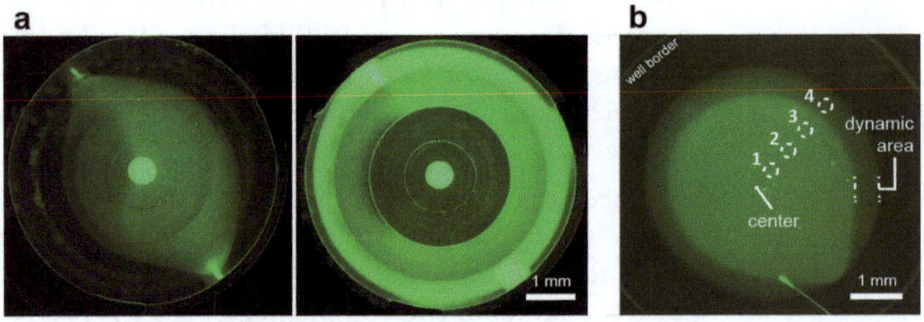

Fig. 3 (**a**) Left: flow confinement covering only three kinetic regions, Right: flow confinement covering all four kinetic regions, (**b**) signal observed on a surface, where regions 1–3 are saturated, while region 4 was incubated for a shorter period of time

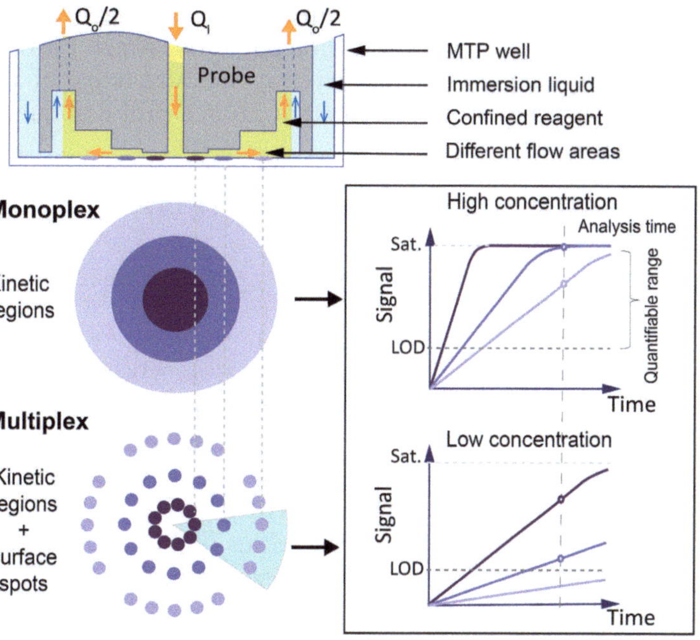

Fig. 4 Types of immunoassays with a probe containing mesas: monoplex, with concentrically organized kinetic regions and multiplex, with additional spots of different analyte types and concentrations

incubation using a 10% of the total assay time is recommended for samples of unknown concentration.

3.3 Data Analysis

3.3.1 Configuration of Kinetic Regions and Arrays of Spots

Mesas can be used to define kinetic regions on any surface with a uniform pattern of a ligand, as shown in Fig. 4. The interest of using such regions is multi-fold: (1) expand the dynamic range of the assay, detect analytes at low concentration below short mesa gaps, where the kinetics are more favorable, and the opposite for high concentration, (2) compare the intensity of multiple regions to approximately infer kinetic information such as k_{on}/k_{off}, and (3) determine signal quality, as shown in the next section.

To calculate the position of mesas and kinetic zones:

1. Consider the number of zones and the difference in binding rate desired. The binding rate is approximately:

$$\dot{b} \approx D\frac{c}{\delta} \tag{2}$$

where D is the diffusion coefficient, c is the concentration of analyte, and δ is the size of the depletion layers. For example, we will aim to obtain a 2 mesa WellProbe with a difference in binding rate of 10× for a common IgG of $D = 6.5 \times 10^{-5}$ mm2 s^{-1}. Thus, the depletion layer under mesa 1 should be 10 times thinner than under mesa 1.

2. Infer the height and central radius of each mesa with the following equation, which provides the thickness of the depletion layer at the outer edge of any mesa of external radius *Ri* and height *Hi* in a probe of number of mesas *n*.

$$\delta = \left(D\frac{\pi}{3Q} \sum_{i=1}^{n} \frac{H_n^3}{H_i} \left(R_i^2 - R_{i-1}^2 \right) \right)^{\frac{1}{3}} \quad (3)$$

where Q is the injection flow rate. For example, for $Q = 40\mu\text{L}\,\text{min}^{-1}$, if $R_1 = 1$ mm and $R_2 = 2$ mm, the height of mesas 1 and 2 should be 0.5 mm and approximately 5 mm, respectively ($\delta_1 \approx 29\mu\text{m}$ and $\delta_2 \approx 321\mu\text{m}$). This equation considers the depletion layer at the outer edge of the mesas, which is usually a good approximation of the depletion layer under the whole mesa. More accurately, however, the central radius of the mesa can be considered. For this, simply substitute the radius of the external mesa in the equation. For example, when calculating δ_2, replace R_2 by $(R_2 + R_1)/2$.

3. In whole surface immunoassays, we can simply measure the values in that position in different regions of the well (all at the right distance R from the inlet).

4. When using antibody spots, the spots should be placed under each mesa at the right distance from the inlet R (Fig. 4). For each identical condition, it is necessary to include a sufficient number of spots, typically at least 3. By placing the same ligand with the same concentration under different mesas, it is possible to obtain a large quantifiable range. Because such configurations allow for a large number of spots, this quantifiable range can be further enlarged by including spots with different analyte concentration. Furthermore, it is possible to test many different analytes, with the space under the inner mesas typically becoming the limiting factor. We recommend a spot diameter of at least 50 μm to ensure sufficient area to accurately average fluorescent signal, although this largely depends on spotting quality.

3.3.2 Reading of Results

Whole surface immunoassay:

1. Image the wells with a microscope and the appropriate fluorescent filter depending on the used fluorophore. Objectives of 2× magnification are usually adequate to visualize all kinetic areas. An exposure time between 200 ms and 3 s is usually recommended, depending on signal level. We recommend taking images of the wells in a wellplate with an automated microscope stage.

2. Using imaging software (e.g., ImageJ/Fiji) or a dedicated application with a programming language (e.g., Python), select

3–4 regions in each kinetic zone and measure average intensity within. Make sure to split color channels if the image is not black and white, and use the most appropriate image (e.g., green channel for FITC). In most experiments, the three zones should be clearly visible. Use this reference for wells where the signal is low or not different between zones. Discard wells with artifacts, such as empty circles due to bubbles or signal outside the area of flow confinement (next to the walls of the well).

3. Average the signal of the three areas for each well, and then average the signal of equivalent wells. Take the standard deviation from the latter.

Array immunoassay:

1. Follow the recommendations of the whole surface immunoassay for image capture. Increase the exposure time if the signal is low. A higher magnification procedure followed by image stitching may be helpful for low-analyte samples.
2. In this case, select a circle inside every spot of the array, covering approximately 80% of its surface. A script is recommended to do this automatically, selecting the central position of 1, 2, or 3 spots, so that the rest are readily identified.
3. Average the signal of equivalent spots. We recommend having wells with replicates and average between them as well as obtaining the standard deviation.

3.3.3 Assessment of Assay Quality

The three-signal zones obtained with WellProbe can be used to assess if the signal is quantifiable and free of artifacts (Fig. 5):

1. In a well-functioning assay, the signal increases from zone 3 to 1 (with zones referring to the areas processed by mesas

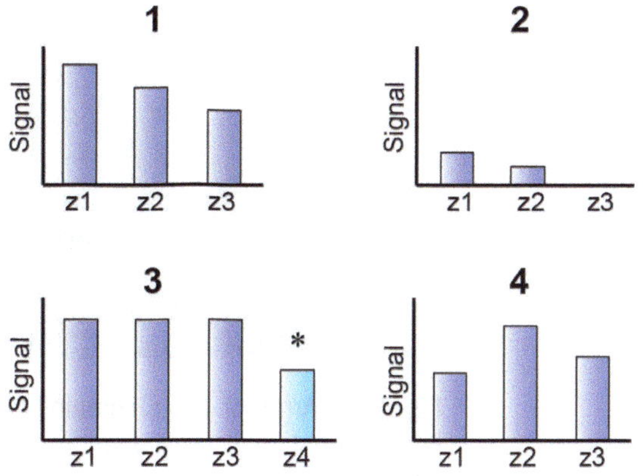

Fig. 5 Four scenarios of signal measured in three concentric zones, after processing by three concentric mesas

3 to 1). The assay is within its dynamic range and is transport-limited, thus benefiting from using WellProbe.

2. The signals can be low and close to their LOD. Depending on the obtained signal strength, zones 1 and/or 2 may still be used for quantitation. Often an increase in flow intensity would enhance the signal.

3. Equivalent high signal intensities in all regions. In this case, saturation has occurred or the chemical system is not transport-limited. In the first case, quantitation is not possible, as no longer in a range of linear relation between time exposure/analyte concentration and signal. In the second case, either the chemical system is not well suited for WellProbe (low transport-dependency, such as for small molecules), or lower flow rates could be used, saving sample. In some cases, reaching saturation may nonetheless be unavoidable, particularly when multiplexing the detection of analytes at very different concentration ranges. In this case, our system allows for the use of a fourth zone, as explained below.

4. Any pattern not corresponding to cases 1–4 would be considered an artifact, indicating that assay quality is low. This could be due to several reasons, from defects in the surface patterning of the ligand, through problems with the flow during the experiment to artifacts in sample reading.

3.3.4 Data Analysis of Spot Arrays

1. Consider only the fluorescent intensity of spots within the quantitation range if the intensity level is more than three times the standard deviation (as measured from identical spots within a well) from the background intensity and less than 3 times the standard deviation from the saturation level (reached in zone 1 if the configuration 3 of Fig. 5). In the example in Fig. 6, the average intensity for each IgG subtype

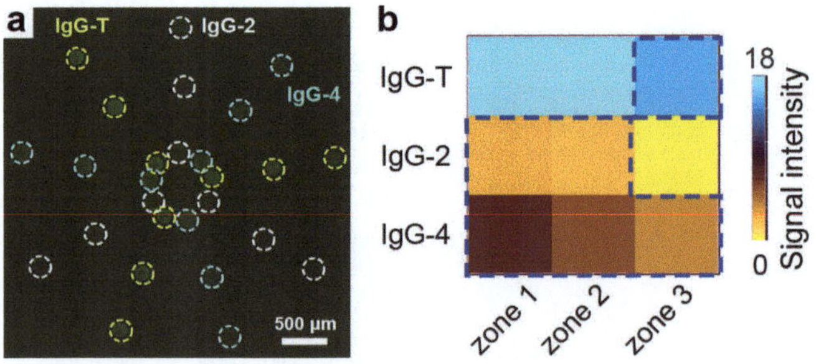

Fig. 6 (**a**) Array of spots containing total IgG and IgG subclasses 1 and 2; (**b**) heatmap obtained from a single array of spots in a well, where each square corresponds to the average of an analytical condition (from 3 spots). Signals in the quantitative range are marked in green

and zone have been illustrated in a heatmap format, and the measurable spots surrounded with a green dotted line.

2. To estimate the quantitation accuracy of the system, use all previously selected values and, assuming a linear behavior, normalize with respect to the results of one condition (e.g., $c = 10$ mg.mL^{-1}, $t = 1$ min) and multiply by a factor corresponding to their difference in dilution or time: e.g., $c = 1$ mg.mL^{-1}, $t = 5$ would be multiplied by $10/5 = 2$. The obtained values should fall within a range of $\pm 50\%$ confidence interval. Discard values outside of this range.

4 Notes

1. For quantification of analyte concentration, we recommend performing a calibration curve using WellProbe. The use of calibration curves based on bench systems may result in inaccuracies.
2. The width of the mesa determines the size and the maximum amount of spots that can be patterned. Due to the geometry of WellProbe, a circular organization of the spots is recommended to have a comparable shear stress on each spot.

References

1. Pereiro I, Aubert J, Kaigala GV (2021) Microscale technologies propel biology and medicine. Biomicrofluidics 15:021302
2. Pereiro I, Fomitcheva-Khartchenko A, Kaigala GV (2020) Shake it or shrink it: mass transport and kinetics in surface bioassays using agitation and microfluidics. Anal Chem 92:10187–10195
3. Serra M, Ferraro D, Pereiro I, Viovy J-L, Descroix S (2017) The power of solid supports in multiphase and droplet-based microfluidics: towards clinical applications. Lab Chip 17: 3979–3999
4. Squires TM, Messinger RJ, Manalis SR (2008) Making it stick: convection, reaction and diffusion in surface-based biosensors. Nat Biotechnol 26:417–426
5. Pereiro I, Cors JF, Pané S, Nelson BJ, Kaigala GV (2019) Underpinning transport phenomena for the patterning of biomolecules. Chem Soc Rev 48:1236–1254
6. Sanjay ST, Dou M, Sun J, Li X (2016) A paper/polymer hybrid microfluidic microplate for rapid quantitative detection of multiple disease biomarkers. Sci Rep-uk 6:30474
7. Pereiro I, Khartchenko AF, Lovchik RD, Kaigala GV (2021) Advection-enhanced kinetics in microtiter plates for improved surface assay quantitation and multiplexing capabilities. Angewandte Chemie Int Ed 60:20935–20942
8. Pereiro I, Khartchenko AF, Lovchik RD, Kaigala GV (2022) Simple add-on devices to enhance the efficacy of conventional surface immunoassays implemented on standard labware. Analyst 147:2040–2047
9. Pereiro I, Khartchenko AF, Petrini L, Kaigala GV (2019) Nip the bubble in the bud: a guide to avoid gas nucleation in microfluidics. Lab Chip 19:2296–2314

Chapter 7

Droplet Microfluidics with Capillary Electrophoresis for Glycan Biomarker Analysis

Théo Liénard--Mayor, Camille Bricteux, Nguyet-Thuy Tran, Arnaud Bruneel, Myriam Taverna, and Thanh Duc Mai

Abstract

Several glycoproteins are validated biomarkers of various diseases such as cancer, cardiovascular diseases, chronic alcohol abuse, or congenital disorders of glycosylation (CDG). In particular, CDG represent a group of more than 150 inherited diseases with varied symptoms affecting multiple organs. The distribution of glycans from target glycoprotein(s) can be used to extract information to help the diagnosis and possibly differentiate subtypes of CDG. Indeed, depending on the glycans and the proteins to which they are attached, glycans can play a very broad range of roles in both physical and biological properties of glycoproteins. For glycans in general, capillary electrophoresis with laser-induced fluorescence detection (CE-LIF) has become a staple. Analysis of glycans with CE-LIF requires several sample preparation steps, including release of glycans from the target glycoprotein, fluorescent labeling of glycans, and purification of labeled glycans. Here, we describe the protocol for glycan sample treatment in a microfluidic droplet system prior to CE-LIF of labeled glycans. The microfluidic droplet approach offers full automation, sample, and reagent volume reduction and elimination of contamination from external environment.

Key words Droplet microfluidics, Capillary electrophoresis, Glycans, Glycoproteins, Magnetic beads, Sample treatment

1 Introduction

Glycoproteins represent half of all secretory and human cellular proteins and play a broad range of functions depending on their glycosylation (protein folding and stability, cellular interactions). Altered glycosylation (i.e., unoccupied glycosylation sites, truncated glycan antennae, and change in proportion of glycoforms) is a posttranslational modification that can affect the regular functioning and activity of glycoproteins [1–3]. Several glycoproteins have been identified as potential disease biomarkers and a few of them as validated biomarkers of various diseases such as cancer, cardiovascular diseases [4], chronic alcohol abuse [5], or congenital disorders of glycosylation (CDG) [6]. Among these, CDG which

Valérie Taly et al. (eds.), *Microfluidics Diagnostics: Methods and Protocols*, Methods in Molecular Biology, vol. 2804,
https://doi.org/10.1007/978-1-0716-3850-7_7,

are caused by genetic defects in the biosynthesis and attachment of glycoprotein glycans, affecting multiple organs, represent a group of more than 150 inherited diseases [7]. No specific treatment exists for CDG, and symptoms are treated to only facilitate the life of the patient. Diagnosis of CDG is thus challenging but crucial and usually requires a whole arsenal of analytical techniques that bring complementary information.

There are several approaches to analyze protein glycosylation at different levels: intact glycoprotein, glycopeptides after protease digestion, glycans after their enzymatical release, and monosaccharides from the glycoprotein. Glycan analysis with capillary electrophoresis with laser-induced fluorescence detection (CE-LIF), which is now an established method for such purpose [7–9], requires several sample preparation steps, including release of glycans from the target glycoprotein, fluorescent labeling of the released glycans and purification of the labeled glycans. One of the conventional methods for in-tube glycan sample treatment prior to CE-LIF uses functionalized magnetic beads as the solid phase extraction support to transfer glycans from one step to another [10, 11]. This process performed in tubes is rather lengthy and requires manual handling. Perkin–Elmer developed recently a microchip-CE platform that transposes the in-tube operations into a 96-well microtiter-plate-based protocol to release and label N-glycans prior to their separation using microchip electrophoresis [12].

In a related context, microfluidics has gained much attention for sample processing thanks to its inherent advantages of reduced sample and reagents consumption, and shorter reaction time [13]. This miniature approach till recently has been little exploited for forefront sample processing for CE-LIF of glycans [14, 15]. The droplet microfluidics, an emerging branch of microfluidics that allows creation and manipulation of discrete droplets considered as micro-reactors inside microfluidic channels [16], has gained attention for this purpose [15]. Indeed, adaptation of a sample preparation protocol into a droplet microfluidic format presents many advantages, notably reduction of working volume down to μL/nL ranges, high throughput, no sample dilution, enhanced mixing and mass transfer inside droplets, and elimination of contamination from external environments. Basic operations normally performed with droplets include merging, splitting, and sorting, which in turn serve for acceleration of chemical reactions, extraction of molecules of interest from complex matrices, etc. In particular, the latter can be implemented by suspending functionalized magnetic beads inside a droplet containing the target molecules to fish them and then using magnets to transfer the beads with the trapped target molecules into another droplet [15, 17].

In this chapter, we describe the protocol we developed to extract glycans from a glycoprotein in a microfluidic droplet format for subsequent CE-LIF analysis of labeled and purified glycans. The microfluidic-droplet protocol holds great synergy with the conventional in-tube protocol since the cornerstone of both approaches is the use of functionalized magnetic beads as the glycan transporters between steps.

2 Materials

All chemicals for preparation of buffers are of analytical or reagent grade. All buffers and samples are prepared with Milli-Q water using a Direct-Q 3 UV purification system.

2.1 Microfluidic Extraction in Droplets

1. Fast Glycan Analysis and Labeling kit from Sciex including Standards (malto-oligosaccharides (MD ladder), 8-aminopyrene-1,3,6 trisulfonic-acid (APTS dye)) and magnetic beads.
2. MD ladder solution: MD ladder at 0.1 mg/mL in deionized water.
3. Magnetic beads suspension: 10 mg/mL.
4. Acetonitrile (ACN).
5. Labeling solution: 40 mM APTS in 20% AcOH/1 M $NaBH_3CN$/NP-40 of 9/1/1 v:v ratio.
6. Denaturation solution: SDS, DTT, and NP-40 at v:v ratio of 1/1/5.
7. Digestion solution: NP-40 and PNGase F at v:v ratio of 11/4.
8. Elution solution: water.
9. Standard IgG solution: 10 mg/mL in PBS, or serum diluted 4 times in water.
10. Microfluidic droplet system: Magelia from Inorevia. The system includes a stepper-motor-driven syringe pump with a 250 μL syringe connected to a 3-port selector valve, a magnetic tweezer, two thermal modules, and three optical detectors. By default, the flowrate is set at 150 μL/min.
11. Fluorinated oil containing surfactant from Inorevia.
12. Oil is stored in the system using PCR 0.2 mL tubes. Reagents and samples are stored in MicroAmp fast reaction tubes (8 tubes/strip).

2.2 CE-LIF Analysis

1. Background electrolyte (BGE) solution containing Triethanolamine (TEOA) and citric acid (Cit): TEOA/citric acid (IS of 150 mM, pH 4.75).
2. Sodium hydroxide (NaOH 1M).

3. CE instrument: Beckman Coulter PA800+ system (Sciex, Brea, CA) coupled with a LIF detector ($\lambda_{excitation}$: 488 nm, $\lambda_{emission}$: 520 nm).
4. Capillary: fused silica capillary of 50 μm id and 375 μm od from Polymicro (TSP075375, CM Scientific, Silsden, UK), with the total length (Ltot) of 60 cm, the effective length (Leff) of 50 cm.

3 Methods

3.1 Fluorescent Labeling of Standard Malto-Oligosaccharides in Droplets

1. Fill the syringe and the tubing in the Magelia system with oil.
2. Aspirate 2.5 μL of magnetic bead suspension in the Magelia system and deliver it to the magnetic tweezer.
3. Activate the magnetic tweezer to trap the beads and discard the supernatant to the trash tube (*see* Fig. 1 and **Note 1**).
4. Aspirate 1 μL of MD ladder solution and deliver it to the magnetic tweezer.
5. Suspend the trapped beads in this MD ladder solution droplet (Fig. 1).
6. Deliver this droplet to the outlet extremity of the tubing.
7. Aspirate 7 μL of ACN and merge it with the droplet containing the beads and the ladder solution (*see* Fig. 1 and **Note 2**).
8. Incubate the merged droplet for 5 min (*see* Fig. 1 and **Note 3**).
9. Activate the magnetic tweezer to trap the beads and remove the supernatant (Fig. 1).

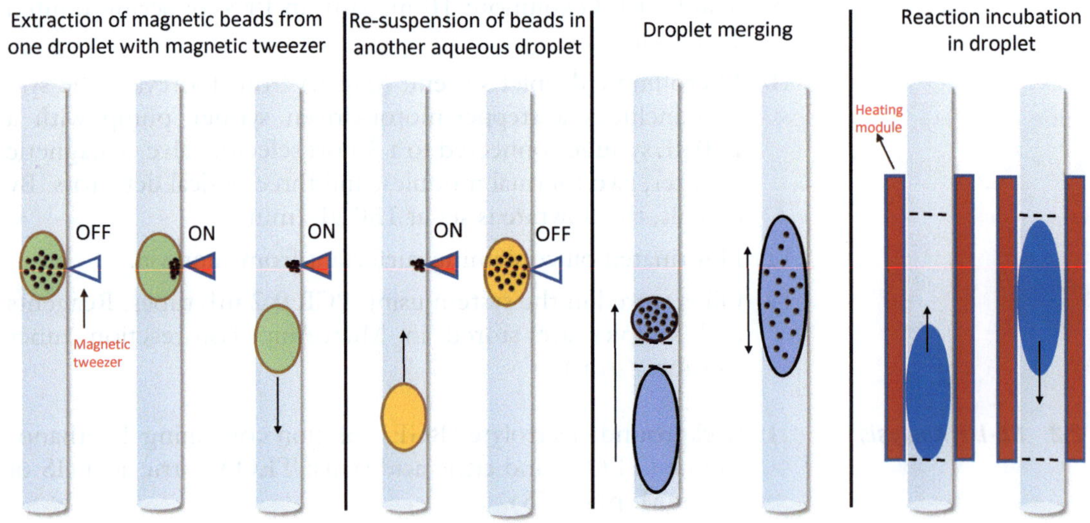

Fig. 1 Basic operations for glycoprotein sample treatment in droplets

10. Aspirate 1 μL droplet of the labeling solution and deliver it to the magnetic tweezer.
11. Resuspend the beads in this labeling droplet.
12. Incubate the mixture for 20 min at 60 °C (*see* **Note 4**).
13. Deliver this droplet to the outlet extremity of the tubing.
14. Aspirate 9 μL of ACN and merge it with the previous droplet containing the beads to jam labeled MD ladder on beads (*see* **Note 5**).
15. Activate the magnetic tweezer to trap the beads and remove the supernatant droplet containing residual APTS to the trash vial (*see* **Note 6**).
16. Aspirate 6 μL water and deliver it to the magnetic tweezer.
17. Suspend the trapped beads into this water droplet for elution of the captured MD ladder.
18. Activate the magnetic tweezer to trap the beads.
19. Collect this eluent from the outlet of Magelia into a 200 μL vial containing oil.
20. Pipette the eluent solution (5 μL out of 6 μL eluent) into an insert and put it into the CE-LIF instrument for subsequent analysis (*see* **Note 7**).

3.2 Denaturation, Digestion, and Fluorescent Labeling of Glycoprotein-Derived-Glycans in Droplets

1. Fill the syringe and the tubing in the Magelia system with oil.
2. Aspirate 2.5 μL of magnetic bead suspension in the Magelia system and deliver it to the magnetic tweezer.
3. Activate the magnetic tweezer to trap the beads and discard the supernatant to the trash tube (*see* **Note 1**).
4. Aspirate 2 μL droplet of standard IgG solution and deliver it to the magnetic tweezer.
5. Suspend the trapped beads in this sample droplet.
6. Deliver this droplet to the outlet extremity of the tubing and merge it with 1 μL of denaturation solution.
7. Incubate this fused droplet (3 μL) over 7 min (*see* **Note 8**).
8. Deliver this droplet (3 μL) to the outlet extremity of the tubing and merge it with 2.4 μL of digestion solution.
9. Incubate this merged droplet (5.4 μL) for 20 min at 60 °C (*see* **Note 9**).
10. Activate the magnetic tweezer to capture the beads after incubation and recover the solution in a vial (*see* **Note 1**).
11. Aspirate 1 μL of the recovered solution and deliver it to the magnetic tweezer.

12. Resuspend the trapped beads in this droplet.
13. Deliver this droplet containing magnetic beads to the outlet extremity of the tubing and merge it with another one of 7 μL ACN (*see* **Note 2**).
14. Incubate the merged droplet over 5 min (*see* **Note 3**).
15. Activate the magnetic tweezer to retain the beads and remove the supernatant.
16. Aspirate 1 μL droplet of the labeling and deliver it to the magnetic tweezer.
17. Resuspend the trapped beads in this labeling solution droplet.
18. Incubate the mixture for 20 min at 60 °C (*see* **Note 4**).
19. Deliver this droplet (1 μL) to the outlet extremity of the tubing.
20. Merge this droplet (1 μL) with 9 μL of ACN to jam labeled glycans on beads.
21. Deliver this merged droplet to the magnetic tweezer.
22. Activate the magnetic tweezer to retain the beads with trapped labeled glycans.
23. Move the supernatant droplet containing residual APTS to the trash vial (*see* **Note 6**).
24. Aspirate 6 μL water and deliver it to the magnetic tweezer.
25. Suspend the trapped beads into this water droplet for elution of the captured glycans.
26. Activate the magnetic tweezer to trap the beads.
27. Collect this eluent from the outlet of Magelia into a 200 μL vial containing oil.
28. Pipette the eluent solution (5 μL out of 6 μL eluent) into an insert and put it into the CE-LIF instrument for subsequent analysis (*see* **Note 7**).

3.3 CE-LIF of APTS-Glycans and Oligosaccharides

1. Precondition the fused silica capillary prior to first use by rinsing with water for 5 min, 0.1 M NaOH for 5 min, water for 5 min, and the BGE for 30 min (*see* **Note 10**).
2. Inject a plug of sample (APTS-labeled malto-oligosaccharide standards or glycans, *see* Subheadings 3.1 and 3.2, respectively) from the inlet of the capillary (10 s at 0.7 psi).
3. Apply a voltage of −25 kV for 20 min (*see* Fig. 2 and **Note 11**).
4. Rinse the capillary with BGE for 5 min after each analysis (*see* **Note 12**).

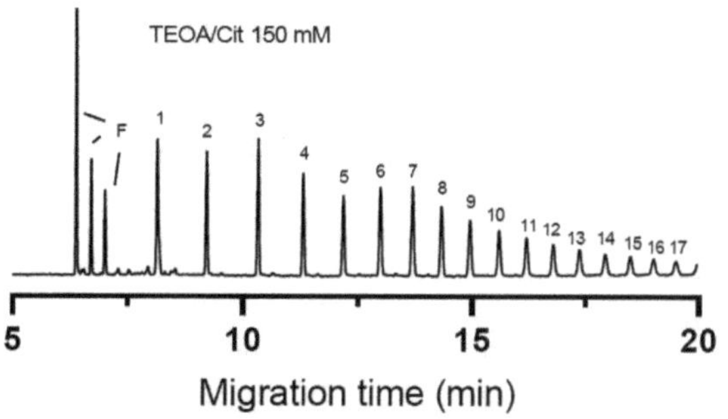

Fig. 2 CE-LIF of oligosaccharides (denoted 1–17). F denotes residual fluorescent peaks. CE conditions: fused silica capillary, 50 μm i.d., 50/60 cm, hydrodynamic injection at 50 mbar for 10 s, HV of −25 kV; BGE composed of TEOA / citric acid (Cit) at ionic strength of 150 mM (TEOA 150 mM and Cit 58 mM)

4 Notes

1. The activation/deactivation of the magnetic tweezer (Fig. 1) is synchronized with the optical detection of droplet passage in the tubing, in order to extract and resuspend the magnetic beads into droplets properly.
2. The merging (Fig. 1) is based on the difference in viscosity of the two droplets. So, the initial distance between two droplets should be sufficiently short (separated with 0.5 μL oil) and the travel distance in the tubing should be sufficiently long (displacement volume of 10 μL). The droplet speed is set at 800 μL/min for the merging.
3. The droplet speed is set at 600 μL/min if incubation (Fig. 1) is implemented at room temperature, in order to allow good circulation of magnetic beads inside the droplet, without breaking the droplet.
4. At elevated incubation temperature (60 °C), the droplet speed is reduced to 300 μL/min to avoid droplet breaking during its movement in the tubing.
5. The maximum droplet volume that the Magelia system can handle correctly is 10 μL. The ACN droplet volume therefore should be calculated to assure that the ACN/total volume ratio reaches at least 0.87 without exceeding the total volume of 10 μL.
6. The repetition of washing step should be optimized (normally two times) to remove well the residual fluorophore without losing much the labeled glycans.

7. Extreme care should be given for collection of the labeled glycan sample from the eluent droplet, to avoid the presence of oil in the collected sample. If oil is accidentally present in the sample to be injected into the CE capillary, this can cause fluctuation or even rupture of the electric current during CE analysis.
8. The droplet speed is set at 600 μL/min.
9. The droplet speed is set at 600 μL/min. Even though the temperature is set to 60 °C, the droplet volume is sufficiently small to avoid droplet breaking at this speed.
10. The analysis of APTS-labeled malto-oligosaccharide standards or glycans (Fig. 2) is carried out with the BGE composed of TEOA / citric acid (IS of 150 mM, pH 4.75) (*see* [18]).
11. No neutral capillary coating is required because the BGE at a high ionic strength allows to suppress efficiently and reproducibly the electro-osmotic flow (EOF). The labeled glycans are separated as anions with the reversed polarity mode under the EOF suppression regime.
12. Inter-run capillary rinsing is done with the BGE for 5 min.

Acknowledgments

This work was supported by the Agence Nationale de la Recherche (ANR, France) with grant no. ANR-18-CE29-0005-01. We thank Inorevia team, in particular Amel Bendali, Julien Autebert, David Gosselin, and Camille Soucies, for their support for instrument development and method optimization.

References

1. Cylwik B, Naklicki M, Chrostek L, Gruszewska E (2013) Congenital disorders of glycosylation. Part I. Defects of protein N-glycosylation. Acta Biochimica Polonica 60:151–161
2. Hennet T, Cabalzar J (2015) Congenital disorders of glycosylation: a concise chart of glycocalyx dysfunction. Trends Biochem Sci 40: 377–384
3. Corfield AP, Berry M (2015) Glycan variation and evolution in the eukaryotes. Trends Biochem Sci 40:351–359
4. Hu MH, Lan Y, Lu A et al (2019) Glycan-based biomarkers for diagnosis of cancers and other diseases: past, present, and future. In: Zhang L (ed) Progress in Molecular Biology and Translational Science, pp 1–24
5. Kohler I, Augsburger M, Rudaz S, Schappler J (2014) New insights in carbohydrate-deficient transferrin analysis with capillary electrophoresis–mass spectrometry. Forensic Sci Int 243:14–22
6. Bruneel A, Cholet S, Tran-Maignan T et al (2020) CDG biochemical screening: where do we stand? BBA Gen Subj 1864:129652
7. Péanne R, de Lonlay P, Foulquier F et al (2018) Congenital disorders of glycosylation (CDG): quo vadis? Eur J Med Genet 61:643–663
8. Feng H-t, Su M, Rifai FN et al (2017) Parallel analysis and orthogonal identification of N-glycans with different capillary electrophoresis mechanisms. Anal Chim Acta 953:79–86
9. Mechref Y, Muzikar J, Novotny MV (2005) Comprehensive assessment of N-glycans

derived from a murine monoclonal antibody: a case for multimethodological approach. Electrophoresis 26:2034–2046
10. Varadi C, Lew C, Guttman A (2014) Rapid magnetic bead based sample preparation for automated and high throughput N-glycan analysis of therapeutic antibodies. Anal Chem 86:5682–5687
11. Sciex (2018) Fast Glycan Labeling and Analysis: High-Resolution Separation, Quantification and ID, https://sciex.com/content/dam/SCIEX/pdf/tech-notes/all/Fast-Glycan-Labeling-and-Analysis-High-Resolution-Separation-and-Identification-in-Minutes.pdf
12. Perkin Elmer (2019) Rapid analysis of n-glycans on the lab chip GXII touch microchip-CE platform, https://resources.perkinelmer.com/lab-solutions/resources/docs/app_rapid_analysis_n-glycans_on_labchip_gxii_touch.pdf
13. Sackmann EK, Fulton AL, Beebe DJ (2014) The present and future role of microfluidics in biomedical research. Nature 507:181–189
14. Aich U, Lakbub J, Liu A (2016) State-of-the-art technologies for rapid and high-throughput sample preparation and analysis of N-glycans from antibodies. Electrophoresis 37:1468–1488
15. Liénard-Mayor T, Bricteux C, Bendali A et al (2022) Lab-in-droplet: from glycan sample treatment toward diagnostic screening of congenital disorders of glycosylation. Anal Chim Acta 1221:340150
16. Sohrabi S, Kassir N, Keshavarz Moraveji M (2020) Droplet microfluidics: fundamentals and its advanced applications. RSC Adv 10:27560–27574
17. Serra M, Ferraro D, Pereiro I et al (2017) The power of solid supports in multiphase and droplet-based microfluidics: towards clinical applications. Lab Chip 17:3979–3999
18. Liénard-Mayor T, Yang B, Tran N-T et al (2021) High sensitivity capillary electrophoresis with fluorescent detection for glycan mapping. J Chromatogr A 1657:462593

Chapter 8

Microfluidic Cartridge for Bead-Based Affinity Assays

Inês F. Pinto, Veronique Chotteau, and Aman Russom

Abstract

Within the vast field of medical biotechnology, the biopharmaceutical industry is particularly fast-growing and highly competitive, so reducing time and costs associated to process optimization becomes instrumental to ensure speed to market and, consequently, profitability. The manufacturing of biopharmaceutical products, namely, monoclonal antibodies (mAbs), relies mostly on mammalian cell culture processes, which are highly dynamic and, consequently, difficult to optimize. In this context, there is currently an unmet need of analytical methods that can be integrated at-line in a bioreactor, for systematic monitoring and quantification of key metabolites and proteins. Microfluidic-based assays have been extensively and successfully applied in the field of molecular diagnostics; however, this technology remains largely unexplored for Process Analytical Technology (PAT), despite holding great potential for the at-line measurement of different analytes in bioreactor processes, combining low reagent/molecule consumption with assay sensitivity and rapid turnaround times.

Here, the fabrication and handling of a microfluidic cartridge for protein quantification using bead-based affinity assays is described. The device allows geometrical multiplexed immunodetection of specific protein analytes directly from bioreactor samples within 2.5 h and minimal hands-on time. As a proof-of-concept, quantification of Chinese hamster ovary (CHO) host cell proteins (HCP) as key impurities, IgG as product of interest, and lactate dehydrogenase (LDH) as cell viability marker was demonstrated with limits of detection (LoD) in the low ng/mL range. Negligible matrix interference and no cross-reactivity between the different immunoassays on chip were found. The results highlight the potential of the miniaturized analytical method for PAT at reduced cost and complexity in comparison with sophisticated instruments that are currently the state-of-the-art in this context.

Key words Microfluidics, Multiplexing, Immunoassays, Colorimetry, Monoclonal antibodies, Biopharmaceuticals, Process analytical technology

1 Introduction

The optimization, monitoring, and control of cell culture processes for the manufacture of biopharmaceuticals remains a highly challenging task due to the multitude of components that continuously change throughout the cultivation process. The presence or depletion of these components can directly impact the overall performance of the process, and affect the critical quality attributes

Valérie Taly et al. (eds.), *Microfluidics Diagnostics: Methods and Protocols*, Methods in Molecular Biology, vol. 2804,
https://doi.org/10.1007/978-1-0716-3850-7_8,

(CQAs) of the product of interest (PoI) [1]. At-line/on-line measurements in mammalian cell cultures are currently limited to (i) monitoring of process parameters, such as pH, pCO_2, pO_2, and cell density, and (ii) quantification of main metabolism markers, such as glucose, lactate, and ammonia [2]. On the other hand, the measurement of small molecules (e.g., amino acids and sugars) and high molecular weight proteins (e.g., product titer and host cell proteins (HCP)) is mostly performed off-line resorting to conventional analytical techniques, such as capillary electrophoresis [3, 4], high-performance liquid chromatography-mass spectrometry (HPLC-MS) [5] and enzyme-linked immunosorbent assay (ELISA) [6]. These methods are time- and resource-consuming, as well as highly costly, which limits the number of analyses that are performed for a given process. In the context of Process Analytical Technology (PAT), there is currently a need for technologies allowing routine at-line monitoring of specific proteins, metabolites, and CQAs in a simple, rapid, and cost-effective manner [7, 8]. Thus, the development of miniaturized analytical systems [9–11] represents an attractive approach to reduce cost-of-goods and achieve a better understanding and control of bioprocesses, both at upstream and downstream levels.

Here, a microfluidic cartridge to perform miniaturized assays for protein analyses in bioreactor samples is described. The device comprises four interconnected bead-packed columns for multiplexed immunodetection of three key protein targets, namely, IgG (product of interest), Chinese hamster ovary (CHO) HCP (main impurities), and lactate dehydrogenase (LDH) (marker of cell viability), and an internal positive control [12]. Nanoporous agarose beads serve as solid phase for either sandwich-based (IgG, CHO HCP) or competitive-based (LDH) immunoassays on chip. The detection is based on HRP-labeled detection antibodies, pre-incubated with the sample under analysis, which allow to generate a colorimetric signal upon reaction with a blotting TMB substrate. The colorimetric signal develops gradually over time, as dark blue precipitates accumulate on the beads, and can be visually interpreted and easily quantified by low-magnification imaging, using for example a flatbed scanner. Microfluidic cartridges are fabricated in poly(dimethyl)siloxane (PDMS), using standard mold replication techniques, and the experimental setup (Fig. 1) to operate the chips consists of a syringe pump to control the flow of the different solutions required to perform the immunoassays. The fabrication procedure to obtain the microfluidic chips and the steps required for implementation of bead-based immunoassays on chip, including off-chip bead coupling, are described in detail in the following sections.

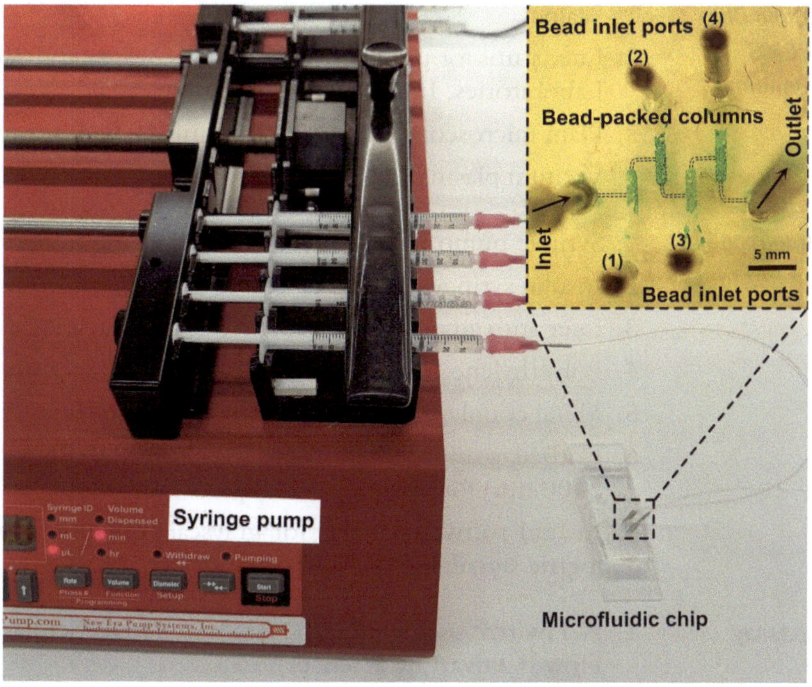

Fig. 1 Microfluidic setup for performing multiplexed detection of protein targets in bioreactor samples. The syringe pump is connected at the inlet of the chip (positive pressure) and a tubing for waste collection is connected at the outlet. Assay parallelization can be achieved by connecting multiple chips to a multi-channel syringe pump. Inset picture: Details of the microfluidic chip design. The bead inlet ports are closed with metal plugs throughout the operation, enabling the liquid to flow through the bead-packed columns. The numbering on the bead inlet ports indicates the sequence for bead packing, according to Table 1. For visualization purposes, a blue dye was flowed through the chip to highlight the bead-packed columns. Dashed lines show the boundaries of the shallow channel interconnecting the bead-packed columns, which allows liquid flow while keeping the beads in place in separate channels

2 Materials

2.1 Microfluidic Chip Production: PDMS Casting

1. Poly(dimethyl)siloxane (PDMS).
2. Curing agent (Sylgard 184, Dow Corning, Midland, MI, USA).
3. Weighing boat.
4. Petri dish.
5. Spatula.
6. Analytical scale.
7. Desiccator.
8. Vacuum pump for PDMS degasification (Chemical Duty Pump, WP6122050, Merck Millipore).

2.2 Microfluidic Chip Production: PDMS Cutting and Bonding

1. Scalpel.
2. Luer stubs for punching access holes (18 ga and 20 ga, Instech Laboratories, Inc.).
3. Plain microscope slides (75 × 25 mm, 0.9–1.1 mm thick).
4. Vacuum plasma system (CUTE-MP/R, Femto Science, Inc.).

2.3 Microfluidic Chip Operation and Readout

1. Syringe pump NE-1002X (New Era Pump Systems, Inc.).
2. Insulin syringes (1 mL U-100 luer lock).
3. Luer stub adapters (20 ga, Instech Laboratories, Inc.).
4. Polyethylene tubing (BTPE-60, Instech Laboratories, Inc.).
5. Metal couplers (SC20/15, Instech Laboratories, Inc.).
6. Flatbed scanner (Epson Perfection V800 Photo, resolution 2400 dpi) for imaging the chips.
7. ImageJ software (NIH) for grayscale quantification of colorimetric signals on the microchannels.

2.4 Immunoassay Components

1. N-Hydroxysuccinimide (NHS)-activated Sepharose 4 Fast Flow (Cytiva Life Sciences).
2. Purified recombinant streptavidin (Thermo Fisher Scientific).
3. L-lactate dehydrogenase from rabbit muscle (Merck, Rahway) for coupling on NHS-activated beads (*see* **Note 1**).
4. Affinity-purified goat anti-CHO HCP (Cygnus Technologies, 3G-0016-AF, Southport) to use as both capture and detection antibody.
5. CHO HCP antigen concentrate (Cygnus Technologies, 3G F553H-1).
6. Goat anti-human IgG Fc preadsorbed antibody (Abcam, ab98616) and goat anti-human IgG H&L preadsorbed antibody (Abcam, ab7148) to use as capture and detection antibody, respectively.
7. Native human IgG protein (Abcam, ab98981) to use as target antigen.
8. Donkey anti-goat IgG H&L preadsorbed antibody (Abcam, ab7120) to use as capture antibody in the internal positive control bead column.
9. EZ-Link NHS-Biotin (Thermo Fisher Scientific) (*see* **Note 2**) and HRP conjugation kit (Abcam) (follow instructions provided by the supplier) for coupling with capture and detection antibodies, respectively.
10. Sample diluent buffer (Cygnus Technologies, I028).
11. Blocker™*CaseininPBS*(*ThermoFisherScientific*).
12. 3,3′,5,5′-tetramethylbenzidine (TMB)-blotting substrate solution (Thermo Fisher Scientific).

3 Methods

3.1 Fabrication of PDMS Microfluidic Chips

The fabrication of microfluidic chips is achieved by soft-lithography based on PDMS replica molding and the procedures should be carried out in a relatively clean environment to minimize trapping of dust particles, such as a dead air box or ideally a laminar flow hood. The mold is a two-level SU-8 mold on a silicon substrate comprising a taller channel (H = 100 μm, W = 700 μm) and a shallower channel (H = 20 μm, W = 200 μm) previously fabricated under class 100 cleanroom conditions.

1. Place the silicon SU-8 mold at the bottom of a transparent polystyrene, heat-resistant Petri dish with the patterned surface facing up. Use tape on each corner of the silicon substrate to fix it to the Petri dish.
2. Weigh 50 g of PDMS pre-polymer and 5 g of curing agent in a plastic weighing boat. Mix thoroughly using a clean spatula until the mixture is visually opaque/milky due to trapped air bubbles.
3. Place the weighing boat in a desiccator and apply a low vacuum for 30 min to remove air bubbles from the PDMS mixture. Note that too high vacuum may cause the liquid to overflow and spill during degassing.
4. Pour the degassed PDMS mixture onto the Petri dish containing the silicon mold and ensure a uniform coverage up to 0.5–1 cm height.
5. Place the Petri dish with the degassed PDMS mixture in a convection oven at 60–70 °C and allow curing for 2 h.
6. Cut the PDMS chip in the silicon mold using a scalpel, allowing sufficient margin around the microchannel features, and peel it off of the mold with the help of tweezers (*see* **Note 3**).
7. Punch access holes using luer stubs (blunt needles) connected to a syringe. Use 18 ga luer stubs for the bead inlet ports and 20 ga luer stubs for the main inlet and outlet (*see* **Note 4**).
8. Clean the PDMS chip (side with the microchannel features) by thoroughly rising with isopropanol and drying with compressed air (*see* **Note 5**). Repeat the same procedure to clean a microscope glass slide with dimensions (width and length) larger than the PDMS chip.
9. Place the clean glass slide and the PDMS chip with microchannel features facing up inside an oxygen plasma chamber.
10. Perform oxygen plasma treatment at 100 W, 600 mTorr O_2 for 30 s.

11. Seal the PDMS chip against the glass slide by putting the two surfaces exposed to the plasma treatment in contact, while pressing uniformly using the tweezers to remove trapped air (*see* **Note 6**).
12. Allow surface stabilization by keeping the chips at room temperature for at least 24 h before use, to approach a plateau of contact angle (*see* **Note 7**). Keep the chips protected from dust by placing them in a closed Petri dish or covering the inlets/outlets with no residue tape.

3.2 Bead Packing in the Microfluidic Chip

Agarose beads (mean particle size, 90 μm) should be coupled with the appropriate capture antibody or antigen prior to packing in the microfluidic chip. Suspend the conjugated beads in a poly-ethylene glycol (PEG) 8000 20% (w/w) solution to facilitate filing in the microfluidic chip (*see* **Note 8**). Prepare a fresh suspension of beads in PEG before each experiment. Avoid air being pushed inside the channels at all stages (*see* **Note 9**).

1. Mount a 1 mL syringe pre-filled with 1× PBS on the syringe pump. Ensure the syringe is free of air bubbles/gaps. Use a luer stub adapter (20 ga) to connect the syringe to 15–20 cm of polyethylene tubing.
2. Connect the free end of the polyethylene tubing to a metal coupler for interfacing with the chip. Prime the tubing by pushing air out using the syringe pump.
3. Connect the metal coupler to the main inlet of the microfluidic chip and start the syringe pump at 5 μL/min in infusion mode (positive pressure).
4. Stop the pump once the liquid starts coming out from the first bead inlet port (closest to the main inlet).
5. Disconnect the metal coupler from the main inlet and close the main inlet using a 20 ga metal plug.
6. Start filling the first column with beads by using a pipette tip at the bead inlet port. Dispense the bead solution slowly using a micropipette and observe the bead packing inside the microchannel. Once the channel is completely filled with beads, remove the pipette tip and close the inlet port using a 20 ga metal plug (*see* **Note 10**).
7. Repeat the previous step for the subsequent columns. Fill one column at a time in a sequential order from the main inlet (Table 1). Ensure that the bead inlet port is properly plugged before moving on to the next column.
8. Remove the plug from the main inlet and connect again the tubing attached to the syringe pump.

Table 1
Sequence of steps for bead packing in the microfluidic chip

	Main Inlet	Bead inlet port 1	Bead inlet port 2	Bead inlet port 3	Bead inlet port 4	Outlet
Load column 1	X	I	O	O	O	O
Load column 2	X	X	I	O	O	O
Load column 3	X	X	X	I	O	O
Load column 4	X	X	X	X	I	O
Load all solutions	I	X	X	X	X	O

I Injection, *O* Open, *X* Closed

9. Start the syringe pump at 5 μL/min in infusion mode and keep flowing until PBS comes out from the main outlet. Ensure the liquid is not leaking from any of the plugged bead inlet ports. Use the packed chip for further processing and sample measurement within 1–2 h.

3.3 On-Chip Bead-Based Immunoassay

For each solution that needs to be flowed through the chip using positive pressure, the tubing needs to be disconnected from the main inlet, filled with the appropriate volume of the next solution, and primed before connecting it to the chip again. Always create an air plug (approx. 0.5–1 cm) in the tubing before aspirating the new solution, to ensure that there is no mixing with the previous solution in the tubing. If the air plug is too large, this may cause a delay in liquid injection due to the need of first compressing a large volume of air.

1. Prepare a small (5–10 cm) tubing with a metal coupler and connect it to the chip outlet. Place the free end of the tubing into an Eppendorf tube to collect waste throughout operation.
2. Fill the inlet tubing with 30–35 μL of casein solution and flow at 2.8 μL/min for 10 min.
3. Fill the inlet tubing with ~55 μL of bioreactor sample (*see* **Note 11**) and flow at 1.7 μL/min for 30 min.
4. Fill the inlet tubing with ~25 μL of 1× PBS, 0.05% Tween, and flow at 5.7 μL/min for 4 min.
5. Replace the inlet tubing connected to the syringe pump by a new one with the same length. Replace also the metal coupler which connects to the chip main inlet (*see* **Note 12**).
6. Prime the new tubing by pushing air out using the syringe pump and leave a 0.5 cm air plug at the end.
7. Fill the inlet tubing with ~30 μL of TMB blotting substrate and flow at 5.7 μL/min for 5 min.

8. Acquire a bright-field image of the chip for colorimetric signal quantification. Wait no more than 2 min after flowing the TMB solution to complete the image acquisition (*see* **Note 13**).

4 Notes

1. Wash a volume of 150 μL of bead slurry with 1.5 mL of cold HCl solution (1 mM) prior to the coupling reaction. Perform this step in a tube using a small bench spinner or a centrifuge. Once the beads are washed, work quickly to avoid hydrolysis of the NHS groups. Prepare streptavidin and LDH at a concentration of 1 mg/mL in an amine-free coupling buffer (e.g., 0.2 M $NaHCO_3$, 0.5 M NaCl, pH 8.3) to maximize the coupling efficiency. Add the protein solution to the washed preactivated beads at a ratio of 1:2 (v/v) (e.g., 10 μL of antibody solution + 20 μL of bead slurry) and incubate at room temperature for 2 h with orbital agitation. Block nonreacted groups on the beads by adding 1.5 mL of 0.1 M Tris–HCl, pH 8.5. Incubate for 1–2 h at room temperature with orbital agitation, followed by incubation with additional 1.5 mL for 24 h at 4 °C. Replace the blocking solution with 1.5 mL of PBS pH 7.4 containing 0.02% sodium azide for long-term storage at 4 °C.
2. Buffer exchange and concentrate the capture antibody to 5 mg/mL in PBS prior to biotin conjugation. This step can be done using Amicon Ultra-0.5 centrifugal filter units with a cutoff of 100 kDa. Prepare the NHS-activated biotin in dimethyl sulfoxide (DMSO) at a concentration of 3.4 mg/mL and quickly add it to the concentrated antibody solution at a ratio of 1:19 (v/v). Incubate in the dark at room temperature for 30 min with continuous agitation. Remove the unreacted biotin by washing 5× with PBS using Amicon Ultra-0.5 centrifugal filter units with a cutoff of 10 kDa.
3. Avoid removing the bulk of cured PDMS from the Petri dish and instead cut only a small area of PDMS corresponding to the chip area. This allows to use less volume of PDMS for subsequent chip replication and prevents the silicon SU-8 mold from cracking due to repeated handling, by keeping it embedded in the bulk of cured PDMS. When cutting the PDMS chips, ensure that the scalpel goes all the way through to the silicon substrate to prevent tearing while peeling them off. The metal scalpel will not scratch the surface of the silicon (Mohs hardness scale of 7), but the material is brittle and excessive force may crack the substrate.

4. Punch the inlets and outlets all the way through by keeping the chip on a cutting mat with the microchannel features facing up. Ensure that a small cylinder-shaped piece of PDMS comes out on the opposite side of the inlet/outlet and use tweezers to remove it completely from the access hole.
5. Wear a lab coat and gloves during chip preparation. Use clean-room tissue paper to clean the surfaces of the PDMS chip and microscope glass slide after rising with isopropanol. Use compressed air to remove any small particles or debris that might remain on the surfaces. Keep the chip and the glass slide protected from dust until plasma treatment.
6. Do not touch the surface of the PDMS chip or glass slide to be bonded, even when wearing gloves, at any point. Even minute amounts of grease on either surface will negatively impact the quality of bonding. Bond the PDMS chip against the glass slide quickly (within 1–2 min) after plasma treatment, since surface activation declines rapidly over time (typically referred to as hydrophobic recovery).
7. Improve the PDMS-glass plasma bonding quality, if needed, by performing an additional bake at 80–90 °C for 15–30 min.
8. Incubate streptavidin-coated agarose beads with each of the biotinylated capture antibodies (e.g., 20 μL of antibody solution at an optimal concentration + 4 μL of bead slurry) in a tube with continuous agitation for 30 min. After incubation, wash the beads with 1 mL of PBS to remove nonconjugated antibodies. Centrifuge the tube to sediment the beads and discard the supernatant solution. Then, resuspend the beads in a PEG 8000 20% (w/w) solution prior to filling in the microfluidic chip.
9. Ensure that the fluidic system—syringes mounted on the pump, tubing, inlet couplers—is free of air bubbles/gaps before starting flowing liquid through the bead-packed channels. Care should be taken when transitioning between different solutions and connecting/disconnecting the metal coupler at the inlet, to prevent air from being pushed inside the inlet port. If an air gap is introduced in the microfluidic channels, this may result in irreversible damage to the bead-packed columns and the chip may need to be replaced.
10. Remove the pipette tip from the bead inlet port without releasing the pipette plunger, to avoid aspirating the beads back into the tip. Leave a small drop of liquid on the inlet port and slowly push the metal plug through the inlet hole until it touches the glass substrate, to ensure that the bead inlet port is tightly sealed for the following operations.
11. Obtain appropriate calibration curves (Fig. 2) for the different immunoassays, by flowing standard solutions of the target

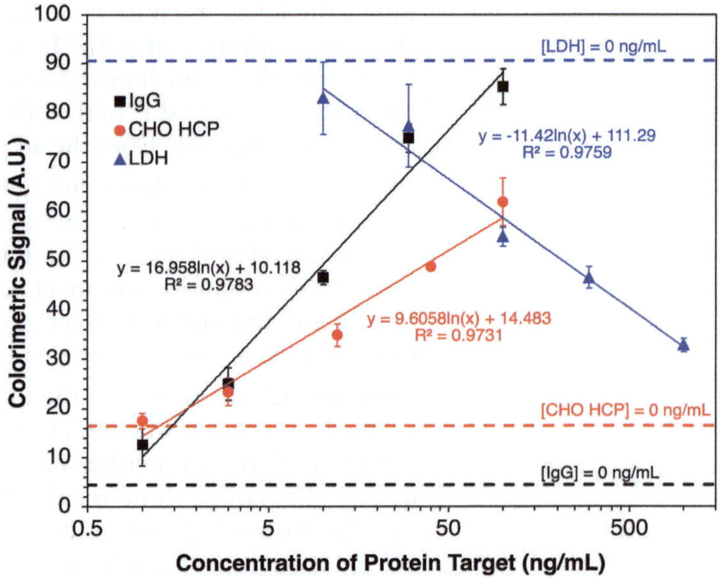

Fig. 2 Concentration curves obtained for (■) IgG, (●) CHO HCP, and (▲) LDH using bead-based immunoassays in the microfluidic chip. Error bars correspond to the standard deviation of three independent measurements of each concentration. Horizontal dashed lines indicate the signals of the negative controls (0 ng/mL target). Limits of detection (LoD) of 0.8 ng/mL, 2.1 ng/mL, and 9.2 ng/mL were obtained for IgG, CHO HCP, and LDH, respectively, considering 3.29σ of the negative control. (Adapted from Ref. [12])

protein at increasing concentrations. Pre-incubate the targets with HRP-labeled detection antibodies at optimized concentrations in a tube at room temperature with continuous agitation. Incubation time can range from 30 min to 2 h, depending on the required assay sensitivity. For the measurement of bioreactor samples, dilute the sample in sample dilution buffer to ensure that the analyte concentration falls within the linear range of the calibration curve. The dilution factor required will be highly dependent on the cell cultivation process under analysis. Prior to loading the sample onto the chip, perform a pre-incubation step with HRP-labeled detection antibodies, as previously described for the calibration curves.

12. Ensure that the fluidic connections are replaced before flowing the blotting TMB substrate, since it rapidly reacts with trace amounts of HRP (from detection antibodies in the sample) that might still be present in the tubing/metal coupler, even after the washing steps. If the TMB reaction starts taking place upstream of the bead-packed columns, the resulting dark-blue precipitates will accumulate on the beads and the real colorimetric signals from the different immunoassays will be masked by high nonspecific background signals.

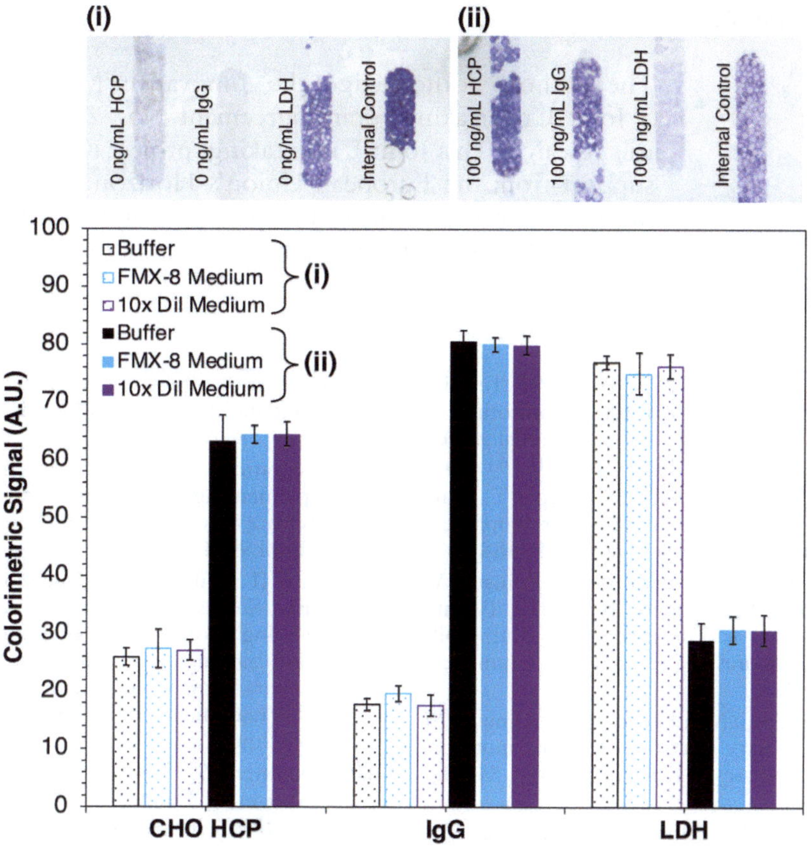

Fig. 3 Microfluidic multiplexed immunoassays demonstrated in different sample matrices, namely, plain buffer, undiluted, and 10x diluted CHO cell culture medium (FMX-8). (i) Samples containing a concentration of 0 ng/mL of all target proteins; (ii) Samples containing [CHO HCP] = [IgG] = 100 ng/mL, [LDH] = 1000 ng/mL. Error bars correspond to the standard deviation of three independent measurements. Images of the bead-packed columns after colorimetric signal development were acquired with a flatbed scanner and are shown above the plot. (Adapted from Ref. [12])

13. Acquire the image of the chip for colorimetric signal measurement always at a constant time-point after flowing the TMB blotting substrate, to ensure consistency of results between different chips. Do not wait longer than 2 min to acquire the images, since the colorimetric signal develops over time, even after stopping the flow of TMB, until it reaches a plateau that will prevent the accurate quantification of the different immunoassays. The intensity of the colorimetric signal on the beads is either directly (IgG, CHO HCP) or inversely (LDH) proportional to the concentration of targets in the bioreactor sample (Fig. 3).

Acknowledgments

The authors acknowledge the Innovative Medicines Initiative 2 Joint Undertaking [grant agreement No 777397] for funding this research. This Joint Undertaking, project iConsensus, receives support from the European Union's Horizon 2020 research and innovation program and EFPIA partners Sanofi, GSK, Bayer, Rentschler, UCB, Byondis, and Pfizer.

References

1. Pais DAM, Carrondo MJT, Alves PM, Teixeira AP (2014) Towards real-time monitoring of therapeutic protein quality in mammalian cell processes. Curr Opin Biotechnol 30:161–167
2. Biechele P, Busse C, Solle D, Scheper T, Reardon K (2015) Sensor systems for bioprocess monitoring. Eng Life Sci 15(5):469–488
3. Ta HY, Collin F, Perquis L, Poinsot V, Ong-Meang V, Couderc F (2021) Twenty years of amino acid determination using capillary electrophoresis: a review. Anal Chim Acta 1174:338233
4. Alhusban AA, Breadmore MC, Guijt RM (2013) Capillary electrophoresis for monitoring bioprocesses. Electrophoresis 34(11): 1465–1482
5. Albrecht S, Kaisermayer C, Reinhart D, Ambrose M, Kunert R, Lindeberg A et al (2018) Multiple reaction monitoring targeted LC-MS analysis of potential cell death marker proteins for increased bioprocess control. Anal Bioanal Chem 410(13):3197–3207
6. Yuk IH, Nishihara J, Walker D Jr, Huang E, Gunawan F, Subramanian J et al (2015) More similar than different: host cell protein production using three null CHO cell lines. Biotechnol Bioeng 112(10):2068–2083
7. Wasalathanthri DP, Rehmann MS, Song Y, Gu Y, Mi L, Shao C et al (2020) Technology outlook for real-time quality attribute and process parameter monitoring in biopharmaceutical development—a review. Biotechnol Bioeng 117(10):3182–3198
8. Gomes J, Chopda VR, Rathore AS (2015) Integrating systems analysis and control for implementing process analytical technology in bioprocess development. J Chem Technol Biotechnol 90(4):583–589
9. Pinto IF, Mikkonen S, Josefsson L, Mäkinen MEL, Soares RRG, Russom A et al (2021) Knowing more from less: miniaturization of ligand-binding assays and electrophoresis as new paradigms for at-line monitoring and control of mammalian cell bioprocesses. Curr Opin Biotechnol 71:55–64
10. Mikkonen S, Josefsson L, Mäkinen ME, Chotteau V, Emmer Å (2022) Capillary and microchip electrophoresis method development for amino acid monitoring during biopharmaceutical cultivation. Biotechnol J 17(8): e2100325
11. van der Burg D, Josefsson L, Mikkonen S, Chotteau V, Emmer Å, Wätzig H et al (2022) Method development for mono- and disaccharides monitoring in cell culture medium by capillary and microchip electrophoresis. Electrophoresis 43(9–10):922–929
12. Pinto IF, Soares RRG, Mäkinen MEL, Chotteau V, Russom A (2021) Multiplexed microfluidic cartridge for at-line protein monitoring in mammalian cell culture processes for biopharmaceutical production. ACS Sens 6(3): 842–851

Part IV

Microchips for Single Cell Analysis

Chapter 9

A Guide to the Quantitation of Protein Secretion Dynamics at the Single-Cell Level

Nathan Aymerich, Olivia T. M. Bucheli, Kevin Portmann, Klaus Eyer, and Jean Baudry

Abstract

Protein secretion is a key cellular functionality, particularly in immunology, where cells can display large heterogeneity in this crucial activity in addition to binary secretion behavior. However, few methods enable quantitative secretion rate measurements at the single-cell level, and these methods are mostly based on microfluidics systems. Here, we describe such a microfluidic single-cell method for precisely measuring protein secretion rates in detail, building on the published droplet-based microfluidic platform DropMap. We give an updated, detailed guide toward quantifying protein secretion rates, discussing its setup and limitations. We illustrate the protocol on two key immunological analytes, immunoglobulin G, and interferon-γ.

Key words Single-cell analysis, Protein secretion, Secretion rates, Immune cells, Quantification, Droplet microfluidics

1 Introduction

Secreted proteins play a major role in cell functionality, particularly in immunology – immune cells secrete cytokines, chemokines, antibodies, or enzymes that regulate and contribute to the immune response. When studying protein secretion, quantitative single-cell techniques are necessary to decipher and characterize the vast heterogeneity within a cell population and link the secreting cell to the secreted protein, its secreted rate, and associated function. The main limitation of common techniques for single-cell protein measurements, such as ELISpot [1] and cytometry-based methods [2, 3], is that they do not allow a quantitative analysis of protein secretion rates as their main output is only the frequency of cells of interest. However, the secretion rate determines the protein concentration in the cell environment and thus greatly influences functionality and tissue responses. To address this issue, different

Valérie Taly et al. (eds.), *Microfluidics Diagnostics: Methods and Protocols*, Methods in Molecular Biology, vol. 2804,
https://doi.org/10.1007/978-1-0716-3850-7_9,

microfluidic techniques have been developed, based on wells [4], chambers [5], or droplets [6] that enable a more in-depth description of the protein secretion at the single-cell level, including the determination of protein secretion rate. These microfluidics platforms have been applied to numerous studies in the field of immunology, such as cancer immunology [7] or antibody discovery [8]. For a more in-depth discussion of the methods for secretion analysis, we refer to our recent review [9].

Here, we present an updated and optimized single-cell method for quantitative protein secretion measurement using the droplet-based microfluidic platform DropMap [10]. This platform was previously used to measure antibody secretion by B cells over an immune response [11], but the method can be adapted to measure many other secreted proteins detectable by sandwich immunoassay. Indeed, various cells can be encapsulated in microfluidic droplets (water-in-oil emulsion), allowing to study multitudes of immune functionalities in different cells. Furthermore, the system is suited for multiplexing, limited only by the number of fluorescence channels available. This would allow measuring up to five different secreted analytes or cell-surface markers for the same cell in parallel. The measurement time can typically range from 30 min to 12 h, depending on the application and secretion rates. A typical throughput lies around 20,000 individual cells per measurement but can be increased up to 100,000 cells if necessary.

This protocol uses a microfluidic PDMS chip to encapsulate single cells and functionalized magnetic nanoparticles to capture the target analyte. The droplets are then immobilized in a 2D observation chamber, and by applying a magnetic field, the nanoparticles in each droplet align into an elongated object called beadline. These beadlines are modified to bind the chosen secreted analyte. A fluorescent detection reagent, specific to the chosen analyte, is also present in the droplets, and the reagent and thus fluorescence will relocate in a concentration-dependent manner onto the beadline if the analyte is secreted (*see* Fig. 1). The droplets containing secreting cells are imaged over time with a fluorescence microscope. The readout of the assay in each droplet is the ratio of the fluorescence on the beadline against the background fluorescence (fluorescence relocation signal, *see* Fig. 1d), assigned to a precise concentration of an analyte using a calibration curve. Thus, knowing the time interval between different time points, a secretion rate can be attributed to every individual secreting cell.

However, a careful adaption of the assay's sensitivity and analytical range toward the expected secretion range and concentration is needed. For immunoglobulin G (IgG) secretion studies, secretion rates can be accurately determined from 5 to 500 IgG/s (range from 0.5 to 15 nM, 5 measuring points with 8-min interval) using the published protocol [10]. The method has been adapted to other secreted analytes, such as cytokines like IFN-γ. However,

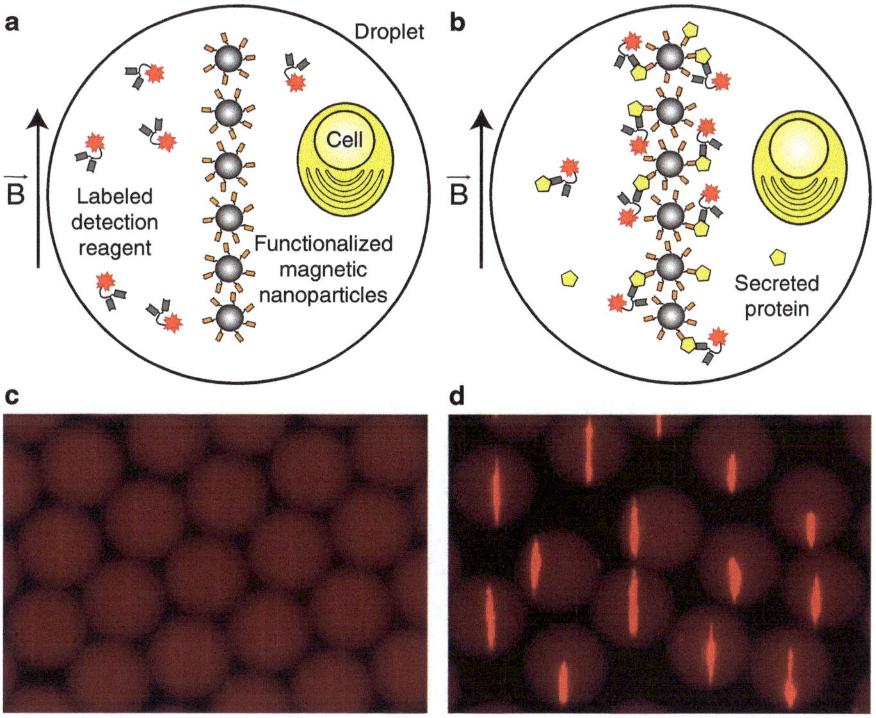

Fig. 1 DropMap fluorescence relocation assay. (**a**) Droplet content: Functionalized magnetic nanoparticles coated with the capture reagent, labeled detection reagent, and a cell. A magnetic field is applied to align the nanoparticles into a beadline. (**b**) The cell secretes the target protein. The molecules are bound by the fluorescent detection reagent and captured on the surface of the nanoparticles, relocating the fluorescence to the beadline. For calibration of the assay, the secreting cell is replaced with a known concentration of target analyte. (**c**, **d**) Images of the IgG application: droplets immobilized in a 2D observation chamber containing calibration samples. (**c**) Droplets containing a blank sample without IgG, no fluorescence relocation. (**d**) Droplets containing a constant concentration of a calibration IgG (10 nM IgG1), fluorescence relocation is visible on the beadlines

for many new analytes, the assay's measurement range, sensitivity, and measurement time need to be adapted to the expected cellular secretion rates to extract reliable information. Therefore, careful adaption and optimization are needed. Here, we describe the necessary procedure for adapting, optimizing, and calculating secretion rates for two analytes, the antibody isotype immunoglobulin G (IgG) and the cytokine interferon-γ (IFN-γ).

2 Materials

Prepare and store reagents at room temperature (RT) unless indicated otherwise.

2.1 Microfluidic Platform and Consumables

1. DropMap droplet generator PDMS microfluidic chip: The fabrication is detailed in Bounab et al. [10], or you can purchase a droplet maker chip capable of producing 50 pL water-in-oil droplets with two inlets for aqueous phases.
2. DropMap glass observation chamber: The fabrication is detailed in Bounab et al. [10], along with magnets and a 3D-printed holder for the chamber and magnets (see https://github.com/ESPCI-LCMD/MiMB).
3. Syringe pump: Three low-pressure modules, such as neMESYS 290 N, and a control module Base 120, CETONI Flow Systems.
4. 1 mL plastic syringes.
5. Needles (23 and 27 gauge).
6. Tweezers.
7. PTFE tubing (0.56 × 1.07 mm and 0.3 × 0.76 mm), Adtech Polymer Engineering.
8. Adapter: 10–32 coned PEEK fitting and ferrule, IDEX.
9. Ring magnet.
10. 200 μL pipette tips.
11. Biopsy punches (0.75 and 6 mm).
12. Fluorinated oil: HFE-7500 3 M™ Novec™.
13. Fluorinated oil with surfactant (outer phase): 0.5% 008-FluoroSurfactant from RAN Biotechnologies in HFE-7500.
14. Mineral oil, light.

2.2 Bioassay Reagents

1. Streptavidin-coated magnetic beads: Bio-Adembeads Streptavidin plus 300 nm, Ademtech, solid content 5 mg/mL. Store at 4 °C.
2. Magnet (neodymium, surface field 2000 Gauss), K&J Magnetics.
3. Phosphate-buffered saline (PBS).
4. D-Biotin, Fluorochem: 0.25 mg/mL in PBS. Store at 4 °C for 3 months maximum.
5. Pan B Cell Isolation Kit II (mouse), Miltenyi Biotec.
6. Capture reagent:

   For the IgG application: CaptureSelect™ Biotin Anti-LC-kappa (Murine) Conjugate, Thermo Fisher. Dilute 1/5 from supplier stock, aliquot, and store at −20 °C.

   For the IFN-γ application: IFN-γ Monoclonal Antibody (MD-1), eBioscience™, 0.5 mg/mL, biotinylated in-house (degree of labeling: 3–5) using EZ-Link NHS-Biotin from Thermo Fisher. Aliquot and store at −20 °C.

7. Pluronic solution: 10% Pluronic™ F-68 in PBS. Aliquot and store at −20 °C.
8. Assay buffer: RPMI 1640 without phenol red, with added 5% KnockOut™ Serum Replacement (Gibco™), 0.5% recombinant human serum albumin (rHSA, *see* **Note 1**), 25 mM HEPES and 0.1% Pluronic™ F-127. Filter sterile, aliquot, and store at −20 °C.
9. Detection reagent:

   For the IgG application: Goat anti-mouse IgG Fc, Alexa Fluor™ 647 conjugate, SouthernBiotech, 1 mg/mL. Aliquot and store at −20 °C (*see* **Note 2**).

   For the IFN-γ application: IFN-γ Monoclonal Antibody (4S. B3), APC, eBioscience™, 0.2 mg/mL. Aliquot, and store at 4 °C.
10. Control antibody:

    For the IgG application: Mouse IgG, any subtype (IgG1, 2a, 2b, 3), 1 mg/mL. For the calibration shown in Fig. 2, we used Mouse IgG1 kappa Isotype Control (P3.6.2.8.1), eBioscience™.

    For the IFN-γ application: Recombinant Human IFN-γ, Peprotech™, 0.17 mg/mL. Aliquot and store at −20 °C.

2.3 Preparation of Cells

1. Centrifuge:

   For the IgG application: Refrigerated to 4 °C.

   For the IFN-γ application: Kept at RT.
2. Cell counter, automated or manual.

2.4 Imaging and Analysis Software

1. Epifluorescence inverted microscope, such as Nikon Eclipse Ti2-E or equivalent, equipped with a heating chamber (gas control is not required), motorized stage, and 10× objective.
2. MATLAB R2020, with *Computer Vision*, *Image Processing*, *Optimization*, *Parallel Computing*, and *Statistics and Machine Learning* toolboxes.
3. DropMap MATLAB application (*see* https://github.com/ESPCI-LCMD/MiMB).
4. Microsoft Excel.

3 Methods

A simple microfluidic platform called DropMap is necessary for this protocol. It comprises a droplet generator PDMS chip, capable of producing ~50 pL water-in-oil droplets, and a glass observation chamber to immobilize the emulsion and image it with time-lapse

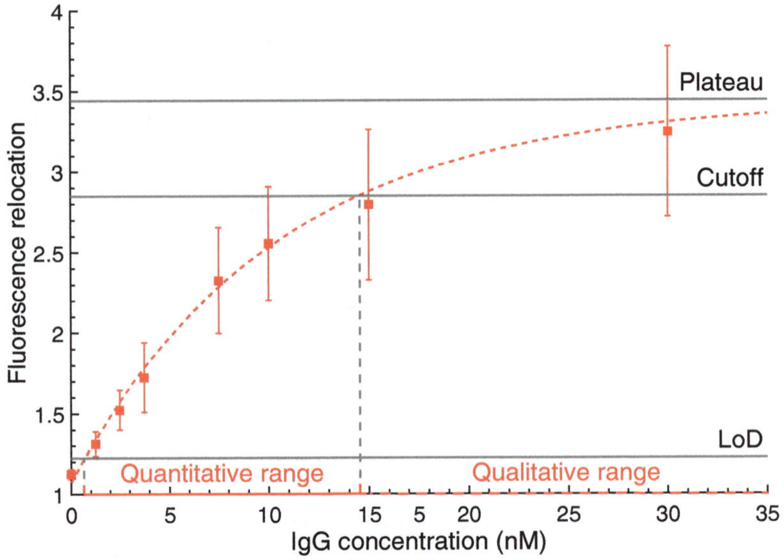

Fig. 2 Calibration curve, IgG application. Median fluorescence relocation values are displayed along with standard deviations and the computed exponential fit (dotted red line). Limit of detection (LoD) and plateau values are also displayed. The cutoff value corresponds to the relocation above which no precise concentration of analyte is computed: due to the variability of the obtained signal, when the relocation is close to the plateau of the calibration curve it is not possible to attribute a precise concentration to a given relocation value. Thus, the assay has a quantitative range, from 0.5 to 15 nM, and a qualitative range, above 15 nM

fluorescence microscopy. The fabrication of chips and observation chambers is described in detail in Bounab et al. [10]. Here, we focus on the preparation of the in-droplet protein secretion assay and its setup, characterization, and optimization to extract precise and reliable secretion rates. As examples, we present two applications: assays to detect and quantify the secretion of either IgG molecules by murine B cells or IFN-γ by T cells from human peripheral blood mononuclear cells (PBMCs). The steps that differ between those two applications are separated into two parts. The protocol can be adapted to various target analytes and cell types by changing the capture and detection reagents, concentrations, calibration, and controls (*see* **Notes 3, 8,** and **11**).

Perform all procedures at RT unless otherwise stated.

3.1 Preparation of Functionalized Magnetic Nanoparticles and Detection Reagents

We present the protocol for a final bead solution volume of 100 μL, sufficient for measuring 4–5 samples (*see* **Note 3**). Scale accordingly for a different final volume (*see* **Note 4**).

1. Resuspend the streptavidin-coated magnetic beads by vortexing. Take 50 μL and transfer to a 1.5 mL tube. Collect the beads at the bottom of the tube using a magnet and discard the supernatant. Resuspend the beads in 100 μL PBS (*see* **Notes 5** and **6**).

2. Add 5 μL of the capture reagent corresponding to the chosen application to the top of the tube (*see* **Note 7**), wash down with the bead solution, and mix by pipetting up and down (*see* **Note 8**). Incubate for 30 min.

   For the IFN-γ application: After incubation, add 1 μL D-Biotin solution to the top of the tube and wash down with the bead solution (*see* **Note 9**). Incubate for 5 min.

3. Collect the beads at the bottom of the tube using a magnet, wait until the supernatant becomes clear, discard it, and resuspend in 100 μL Pluronic solution. Incubate for 30 min.

4. Repeat the procedure in **step 3**, this time with 100 μL of assay buffer.

5. Collect the beads and discard the supernatant.

   For the IgG application: Resuspend in 97.8 μL assay buffer. Add the detection reagent: 2.17 μL IgG Fc probe, corresponding to 75 nM in-droplet concentration (*see* **Notes 10** and **11**).

   For the IFN-γ application: Resuspend in 98.6 μL assay buffer. Add the detection reagent: 1.36 μL IFN-γ Monoclonal Antibody (4S.B3), corresponding to 5.33 nM in-droplet concentration.

   Adjust the volume of assay buffer to have a total volume of 100 μL in case different detection reagents are used.

3.2 Calibration of the Assay and Controls

Before measuring the secretion of individual cells, it is necessary to calibrate the assay by measuring the signal obtained with known concentrations of the analyte inside the droplets. With the obtained calibration curve, it is then possible to link the signal measured in individual droplets to the analyte concentration, yielding a secretion rate in a time-lapse measurement.

1. Prepare a serial dilution of the target analyte in assay buffer, for example:

   For the IgG application: An IgG control of 100 nM is prepared (1.6 μL added to 98.4 μL assay buffer), with five serial dilutions by a factor 2. Also, prepare a blank sample composed of assay buffer only. Thus, the final in-droplet concentrations of the IgG standards will be 50.0, 25.0, 12.5, 6.3, 3.1, 1.6, and 0 nM (*see* **Notes 10** and **12**).

   For the IFN-γ application: Similarly, a blank sample and serial dilutions of recombinant human IFN-γ are prepared to range from 80 nM (1.6 μL added to 98.4 μL assay buffer) to 0.1 nM in-droplet concentration.

2. Successively co-encapsulate the standards with beads and detection reagents using the DropMap system (*see* Subheading 3.4), image the obtained filled chambers, and proceed with image analysis (*see* Subheading 3.5) before going to the next step.

3. For each calibration sample, the software outputs individual droplet data of fluorescence relocation, in one or several fluorescence channels corresponding to the fluorescent detection reagent(s). For each standard, compute the median and median absolute deviation (MAD) of the fluorescence relocation of individual droplets to plot the calibration curve for visualization (*see* **Note 13**). In this example, fluorescence relocation in the Cy5 channel increases with increasing IgG/IFN-γ concentration (*see* Fig. 2). This calibration curve will be used to compute the secretion rate of the target analyte by single cells (*see* Subheading 3.6). Here, we only use one detection reagent for each application, but it is possible to use several detection reagents in parallel. In that case, repeat the procedure in the corresponding fluorescence channels to obtain all the corresponding calibration curves (*see* **Note 14**).

3.3 Preparation of the Cell Suspension

The analyzed cell populations for the IgG secretion application are usually B cells enriched from mice's spleen or bone marrow using the Pan B Cell Isolation Kit II (mouse). For the IFN-γ secretion application, the cells are usually PBMCs isolated from human blood. However, any population of cells from a mouse, human, or other sources can be used to detect the secretion of a given analyte (*see* **Notes 15** and **16**).

1. Determine the cell count per mL using a cell counter. Store the cell suspension on ice (for the IgG application) or at RT (for the IFN-γ application, *see* **Note 17**) until measurement time.
2. Take a volume corresponding to one million cells and transfer the cells into a 1.5 mL tube (*see* **Notes 18** and **19**). Centrifuge (400 g, 5 min at 4 °C for the IgG application or at RT for the IFN-γ application), discard the supernatant, and gently resuspend the cells in 100 μL assay buffer (*see* **Notes 20**, **21**, and **22**). Store the tube on ice (for the IgG application) or at RT (for the IFN-γ application).

3.4 Encapsulation

Once the two aqueous phases are prepared (beads and detection reagents, and either calibration or cell sample), the next step is to co-encapsulate them into 50 pL water-in-oil droplets using the PDMS droplet-generator chip and the syringe pumps. The emulsion is then directly introduced into a 2D glass observation chamber, where a monolayer of droplets is immobilized and imaged over time. The chamber is placed between two magnets applying a uniform magnetic field that aligns the functionalized magnetic nanoparticles into beadlines.

1. Connect a 25 cm long PTFE tubing (0.56 mm inner diameter) to a 200 μL pipette tip using a round PDMS slab (6 mm diameter, ~5 mm thickness, with a 0.75 mm hole in the middle pierced with a biopsy punch, *see* Fig. 3 and **Note 23**).

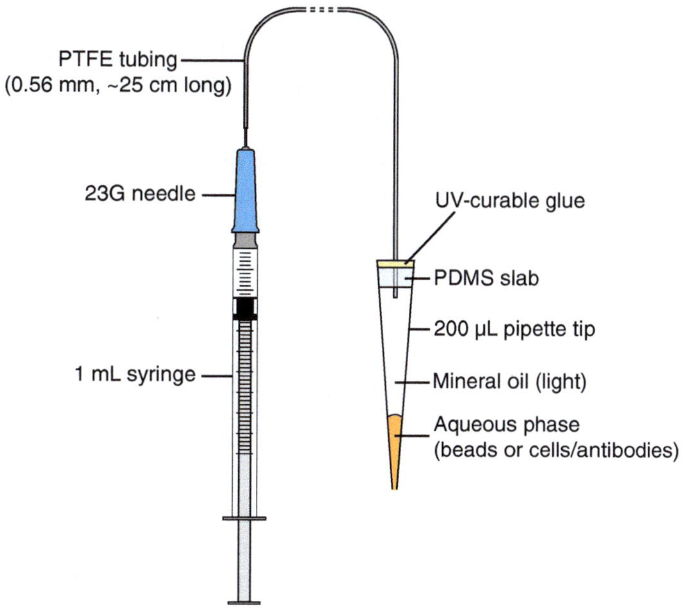

Fig. 3 Assembly for the injection of the aqueous phases into the droplet generator microfluidic chip

2. Use tweezers to connect the other end of the tubing to a 23 gauge needle, and attach it to a syringe (*see* **Note 24**).
3. Fill the assembly with mineral oil (*see* **Notes 25** and **26**).
4. Repeat **steps 1** to **3** to create a second assembly. These will take up the two aqueous phases and inject them into the microfluidic chip.
5. Use tweezers to connect a 25 cm long PTFE tubing (0.3 mm inner diameter) to a 27 gauge needle, and attach it to a syringe (*see* **Note 24**). Fill the assembly with fluorinated oil with surfactant; it will be used to inject the oil phase into the chip (*see* **Note 25**).
6. Install the syringes on the three modules of the syringe pump.
7. To take up the two aqueous phases, place the tips inside the respective tubes, and apply a negative flow rate of 2,000 µL/h to aspirate 30 µL (*see* **Note 27**).
8. Connect the tubing for the oil phase to the outer-phase inlet of the microfluidic chip and apply a flow rate of 2,000 µL/h until oil flows out of the outlet and the aqueous phase inlets, ensuring that all the air inside the channels of the chip has been eliminated, then stop the flow and wipe the excess outflow.
9. Connect the two tips with the aqueous phases to the inner-phase inlets (*see* **Note 28**).

10. Apply a flow rate of 200 μL/h for the two aqueous phases. Wait until they reach the chip outlet, and then start the oil flow with a flow rate of 800 μL/h (*see* **Note 29**). This generates droplets of ~50 pL.
11. As soon as the droplets flow out of the outlet of the chip (visible by turbid and flat outflow), connect a 10 cm long piece of 0.3 mm inner diameter PTFE tubing to the outlet of the chip.
12. Connect the other end of the tubing to the glass observation chamber using the adapter when the droplet flow reaches the end of the tubing (*see* **Notes 30** and **31**). The chamber should fill with a straight front of droplets. The droplets will flow inside the chamber, forming a monolayer (see **Notes 32 and 33**).
13. When the chamber is fully filled with droplets, stop the flow of all the modules, disconnect the adapter, and absorb the excess liquid in each chamber port with a lab wipe before closing it (*see* **Note 34**).

3.5 Imaging and Image Analysis

The imaging of the droplets is done with an epifluorescence microscope, acquiring a large image, typically composed of 10 × 10 fields of view (square size of 1.3 × 1.3 cm with our acquisition setup), in all the fluorescence channels corresponding to the different detection reagents, as well as the brightfield channel. The acquired images are saved as .nd2 or .tiff files.

The chamber is usually imaged every 8 min for 40 min in the red fluorescence and brightfield channel (indicative times) for the IgG application. For the IFN-γ application, the chamber is usually imaged every 30 min for 4 h in the same channels. For different analytes, the measurement's time interval and complete duration should be adjusted depending on the secretion rates and the range of the assay (see **Note 35**). We use a blacked-out thermalized enclosure to maintain the chamber at 37 °C during the measurement (see **Notes 36** and **37**).

The image analysis is performed using a custom MATLAB application called DropMap [10]. Alternatively, any equivalent custom program can be used to detect the droplets in the brightfield channel, save their identity and position, and measure the fluorescence relocation on the beadline for each droplet.

3.6 Data Analysis

After processing the image corresponding to a measurement, the DropMap application outputs two data files, a MATLAB data file (.mat) and an Excel file (.xlsx). The MATLAB file is more comprehensive (*see* **Note 38**), but both data files contain the primary data needed for further analysis: the fluorescence relocation signal in each channel used in the assay for every detected droplet over the imaged time points. This primary relocation data is further analyzed, either with MATLAB or Excel (or another program), following the steps below:

1. Compute an approximate of the real lambda value (mean cell number per droplet) in the measurement: count the number of cells present in each droplet for 300 random droplets and compute the sum N (*see* **Note 39**). An approximate of the lambda value is then: $\lambda \approx N/300$.
2. Compute the limit of detection (LoD) of the analyte in the measurement, corresponding to the signal obtained for the smallest concentration of analyte that can be reliably detected and distinguished from a blank sample (*see* Fig. 2 and **Note 40**). This is done using the population of empty droplets in the measurement at the first time point (*see* **Note 41**):

$$\text{LoD} = \text{median}_{\text{empty droplets}} + 1.645 * 2 * \hat{\sigma}$$

$$\hat{\sigma} = k * \text{MAD}_{\text{empty droplets}}$$

where $\text{MAD}_{\text{empty droplets}}$ is the median absolute deviation in empty droplets, $k = 1.4826$, and k*MAD is a consistent estimator of the standard deviation σ, considering the distribution of relocation signals for empty droplets as normal (*see* **Notes 42** and **43**).
3. Sort droplets according to the following three criteria:

   I. The droplet contains a detectable concentration of analyte: at least one timepoint with a relocation signal above the LoD.

   II. The relocation signal of the analyte increases over time: positive slope over the time series.

   III. The signal varies more over the time series than the analytical noise in the absence of analyte:

$$\max_{\text{relocation signals}} - \min_{\text{relocation signals}} > 1.645 * 2 * \hat{\sigma}$$

   The droplets that fulfill all three criteria are classified as containing a secreting cell (*see* **Notes 44, 45,** and **46**).
4. Fit the calibration curve of the analyte using an exponential model:

$$y = y_{\text{plateau}} - \left(y_{\text{plateau}} - y_{\text{blank}}\right) * e^{-K*x}$$

$$x = -\ln\left(\frac{y_{\text{plateau}} - y}{y_{\text{plateau}} - y_{\text{blank}}}\right)/K$$

where x is the concentration of the analyte, y is the fluorescence relocation, y_{blank} is the median relocation of a blank sample, and $y_{plateau}$ is the fitted plateau value.

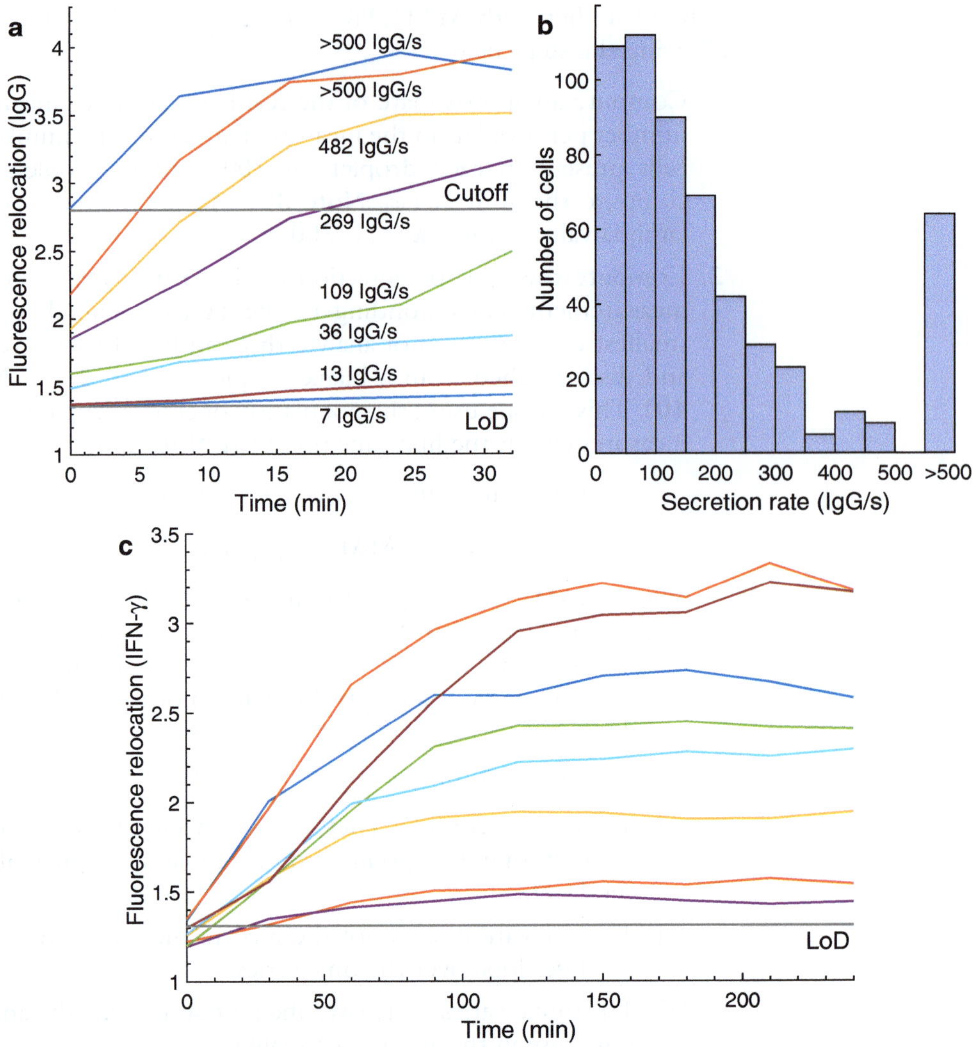

Fig. 4 Results of single-cell secretion measurements. (**a**, **b**) IgG application, measurement of B cells enriched from spleens of mice immunized with PE and a TLR9 agonist (CpG ODN 1826), harvested 6 days after secondary immunization. (**a**) Time-lapse measurement of the fluorescence relocation for selected IgG-secreting cells, along with their mean secretion rates. Precise secretion rates are computed only for the time intervals where both relocation signals are below the cutoff value (*see* **Note 48**). (**b**) Distribution of the mean secretion rates of IgG-secreting cells (pooled values from separate measurements of 6 replicate mice). (**c**) IFN-γ application, measurement of human PBMCs. Time-lapse measurement of the fluorescence relocation for selected IFN-γ-secreting cells

To compute the fit, use individual droplet data for the different concentrations (20,000 droplets each for example).

This fit can be used to attribute a concentration of analyte x to the fluorescence relocation y of a droplet. For every positively sorted droplet, compute the concentration of analyte at each time point using this method and derive a mean secretion rate SR throughout the measurement (*see* Fig. 4, **Notes 47** and **48**):

$$SR = \frac{1}{n-1} \sum_{i=1}^{n-1} \frac{[\text{analyte}]_{i+1} - [\text{analyte}]_i}{\Delta t}$$

where n is the number of time points, Δt is the time interval, and $[analyte]_i$ is the analyte concentration at the time point i.

4 Notes

1. The rHSA can be dissolved directly in RPMI 1640 without phenol red as a 10% w/v solution, aliquoted, and stored at − 20 °C.
2. Aggregates of fluorescent detection reagent can form over prolonged storage and will then be visible in the droplets when imaging, disturbing the assay. To eliminate them, centrifuge the aliquot at 10,000 g for 10 min before using the reagent.
3. When adapting the assay to another analyte and thus using different capture and detection reagents, it is crucial to check that the reagents display the right specificity using appropriate controls. For example, for the IgG application, we verified that the IgG detection reagent binds all IgG subtypes similarly and that it does not crossreact with other immunoglobulins such as immunoglobulin A (IgA) or immunoglobulin M (IgM). This validation can be done using monoclonal control antibodies.
4. Maximum volume tested: 500 μL.
5. The functionalized nanoparticles and detection reagent solution should preferably be prepared fresh on the day of the measurement, but it is also possible to prepare them in advance without adding the detection reagent(s) and store them for up to 2 days in the fridge. On the day of measurement, the solution can be stored at RT during use, but we recommend to store it in the fridge or on ice during gap times in the protocol. Depending on the fluorophore(s) present on the detection reagent(s), the solution should be protected from direct light. Lastly, the beads will sediment over time, and it is necessary to resuspend the solution before use.
6. After collecting the beads and discarding the supernatant, directly resuspend the beads in the new liquid phase. Letting the beads dry up will result in reduced performance.
7. The binding of the biotin conjugate on the streptavidin-coated beads is quick and nonreversible. This is why the drop of biotin conjugate is first added on the top side of the tube before washing down with the full volume of bead solution to ensure a homogeneous binding on the beads, and this handling also serves as a visual control for addition.

8. Adjusting the binding capacity of the beads contained in one droplet: In the IgG application example, the maximum binding capacity of the capture beads is 75 nM (4.5×10^6 IgG molecules in each 50 pL droplet). When adapting the assay to another analyte by using another specific biotinylated capture reagent, this capacity should be adjusted to be greater than the maximum amount of analyte secreted in the duration of the measurement. The capacity can be measured by performing calibration experiments with increasing concentrations of analyte and a fixed concentration of detection reagent: when the relocation signal starts decreasing, the beads' capacity is exceeded. The capacity is proportional to the number of beads and the density of the capture reagent present on them. It is preferable to saturate the beads with the capture reagent; afterward, the capacity can be adjusted by modifying the number of beads per droplet. This protocol results in around 1,200 beads/droplet, but the bead number can be increased up to 2,400 without any limitations.
9. Adding D-Biotin blocks all the non-bound streptavidin present on the beads to reduce aggregation. This step is unnecessary for the IgG application, as the beads do not usually aggregate in that case.
10. Consider that the solution containing the nanoparticles and detection reagents will be co-encapsulated in a ratio of 1:1 (v/v) with the second aqueous phase prepared in Subheadings 3.2 or 3.3. The final concentration of each reagent in the droplets will therefore be half its concentration in the respective aqueous phase. We refer to this as "in-droplet" concentration.
11. Adjusting the concentration(s) of detection reagent(s): The concentration of the detection reagent determines the effective range and sensitivity of the assay; it should be adjusted to match the range of target analyte concentration after secretion by a single cell for the chosen imaging time. For example, in the IgG application, the range of the assay for precise quantification is 0.5 to 15 nM in-droplet concentration (*see* Fig. 2). As precise data are often not available, a rule of thumb to adjust the concentration of the detection reagent is to start with an estimate of the maximum concentration of analyte that should be measurable by the assay (in case of a one-to-one binding of the detection reagent to the analyte; in case of a N-to-one binding multiply this concentration by N; also *see* **Note 12**). This is, however, just a starting point and further optimization should be pursued. Here, for example, the recommended concentration of the detection reagent does not follow this rule for either application.

When performing a calibration curve (*see* Subheading 3.2), the plateau of the curve should be reached at a slightly higher concentration than the desired maximum measurable concentration. If the plateau is reached earlier, the concentration of the detection reagent should be increased. There is also a tradeoff between the lower and higher limits of the effective range of the assay: if the plateau of the calibration curve is reached at a much higher concentration than the expected concentration in most droplets containing a secreting cell, it is preferable to decrease the concentration of detection reagent. This consideration should improve the assay's sensitivity by decreasing the detection limit.

12. The different concentrations of standards should span the expected range of target analyte concentration after secretion by a single cell for the chosen imaging time (30–60 min for the IgG application). If the secretion rates and their range are not known from the literature, start with an estimate. Here, try to measure a typical cell sample and adjust the range of the assay if necessary: if the signal reaches the plateau of the calibration curve in too many droplets, the higher limit of the range of the assay needs to be increased.
13. The blank calibration sample should have no relocation of the detection reagent(s) on the beadline: median fluorescence relocation value(s) close to 1 (usually below 1.15). Otherwise, it shows that there is an unspecific binding of the detection reagent(s) on the beads, which will greatly reduce the sensitivity and measurable range of the assay. This observation can, for example, be due to interactions between the capture and detection reagents. If so, the composition of the assay should be modified to circumvent the issue.
14. Single probe controls are necessary when using several detection reagents to check for potential interactions between the probes that would interfere with the assay. Measure a control sample with a concentration around the middle of the range of the assay (e.g., 10 nM IgG for the IgG application) with each of the probes alone, and compare the different relocations to those obtained with the full assay. There should not be any significant difference in median relocation values for each probe; otherwise, it means that the probes interact with each other and the composition of the assay should be modified.
15. If a detection reagent present in the assay is a full IgG or the analyte is an Ig molecule, perform FcR blockage before cell encapsulation to avoid unspecific binding/activation.
16. Handle the waste properly, according to the local regulations of your institution and government corresponding to the biosafety classification of the samples.

17. For the IFN-γ application, and in general, when measuring the secretion of cytokines by innate immune cells, the cells should be handled at RT because certain immune cells can induce stress signals, such as TNF-α, when handled on ice. We have never observed any bias induced by the ice storage on ice for the IgG application. However, storage on ice improves cell viability over prolonged storage (up to 24 h) and limits unwanted IgG secretion in bulk while the cells are stored.
18. Always pipette the cells using tips with large openings (e.g., 1,000 μL tips or wide-bore 200 μL tips). This will limit the mechanical constraints on the cells caused by pipetting.
19. Extreme care should be taken when handling the cells throughout the protocol. In particular, the cell suspension should always be kept on ice. Stress caused by additional centrifugation steps or temperature changes can decrease the viability and impact the secretion rates of the target analyte(s). When performing a quantitative comparison of secretion rates between several samples, it is crucial to handle all the samples identically to avoid introducing any bias.
20. This leads to a lambda value (mean cell number per droplet) of approximately 0.4. The lambda value can be decreased to limit the probability of having more than one cell in a droplet or increased when searching for very rare events (very low frequency of cells secreting the target analyte in the cell population). The theoretical relationship between the lambda value and the number of cells (N) resuspended in the 100 μL sample (V_{sample}) used for encapsulation is as follows:

$$\lambda = 0.5 * V_{droplet} * N/V_{sample}$$

with $V_{droplet} = 50$ pL. In practice, the obtained lambda value can deviate slightly from the target value; if a systematic bias is observed, adjust the starting number of cells accordingly.
21. If an important secretion of the analyte is expected, it is possible to wash the pellet before resuspending to lower the background signal: after discarding the supernatant, add 100 μL of assay buffer without disturbing the pellet, remove the 100 μL and discard, and then resuspend the cells properly in 100 μL of assay buffer.
22. The cells should be well individualized in the suspension for successful single-cell encapsulation.
23. Option: After inserting the tubing into the PDMS slab and fitting the slab inside the 200 μL pipette tip, it is possible to add UV-curable glue on top to fix the assembly permanently (*see* Fig. 3).

24. Use sterile needles. Be very careful when inserting the needle into the tubing. It is absolutely necessary to hold the tubing with tweezers and not directly with your fingers to prevent injury.

25. There should not be any air inside the assembly. To get rid of potential air bubbles, hold the assembly upward and push until the bubbles flow out of the opening. Refill the syringe if necessary.

26. HFE-7500 can be used here as an alternative to mineral oil.

27. For IgG secretion application: The B cell suspension prepared in Subheading 3.3, **step 2** should be kept on ice while taking up the necessary volume for encapsulation with the syringe pump. Afterward, it is necessary to work quickly during the rest of the encapsulation procedure. The goal is to limit the secretion of the analyte as much as possible before the cells are encapsulated. Otherwise, a considerable amount of the analyte in the solution will lead to a background signal in every droplet, decreasing the assay's sensitivity. If necessary, the cells can be cooled during the encapsulation process by using a 2 mL tube with a hole pierced at the bottom through which the tip used to inject the cell suspension is inserted. The 2 mL tube is then filled with ice-cold water.

    For cytokine secretion applications: Handle the cells at RT (*see* **Note 17**).

28. Usually, a microfluidic droplet maker chip with separate entries for the two aqueous phases is used, one for the beads and detection reagents and one for the sample. This allows the mixing of the two aqueous phases directly inside the droplets. Using a chip with only one entry for the aqueous phase is also possible. This chip design would then require manual mixing of the two aqueous phases in advance. In that case, care should be taken to limit the secretion of the analyte before the injection. Keep both aqueous phases on ice beforehand and form the droplets as quickly as possible.

29. These flow rates correspond to the droplet maker chip described in Bounab, Eyer et al. [10]. With a different chip design, the flow rates should be adjusted to obtain 50 pL droplets.

30. Adding a ring magnet around the tubing between the droplet generator chip and the observation chamber is preferable. The magnetic field applied during the transfer facilitates the beadline formation, an effective solution when multiple thin beadlines are observed in each droplet.

31. It can be preferable to eliminate the first droplets that flow through, for example, by directing the flow into a lab wipe for a few seconds before connecting the tubing to the chamber because the emulsion can be heterodisperse at first before stabilization.
32. While filling the chamber, leave the other end open, placing a piece of lab wipe inside to collect the excess flow.
33. If the emulsion is homogeneous and monodisperse (only well-formed ~50 pL droplets), the array of droplets inside the chamber will look like a homogeneous turbid phase to the naked eye while filling. Larger droplets or air bubbles will be visible, and the flow should be continued until they flow out of the chamber. To eliminate air bubbles faster, it helps to tilt the chamber with the open end upward.
34. Do not overtighten the plugs for the chamber's ports, it will damage the ports and plugs over time, and the chamber will lose its hermeticity.
35. For measurements longer than 2 h, the movement of droplets might become problematic. In that case, the concentration of surfactant in the fluorinated oil can be increased to 2% to improve droplet stability.
36. It is also possible to image the chamber at RT. The decreased temperature will typically reduce the secretion rate of the cells. For IgG secretion measurements of B cells, we have not observed any other apparent change in the characteristics of the secreting cell population compared to measurements at 37 °C. However, for certain cytokines (for example, IL-6), no secretion is observed at RT; therefore, the measurement at 37 °C is mandatory.
37. When measuring at 37 °C, let the chamber equilibrate for 2–5 min before starting the measurement. This step allows all the solutions to heat up and prevents the movement of droplets during image acquisition.
38. Even when performing the analysis with Excel, the MATLAB file should be kept, as it is mandatory to use the *Verify & Sort* function of the DropMap application.
39. Counting the number of cells in more than 300 droplets will increase precision, and it is necessary to do so for very low lambda values ($\lambda < 0.1$). For intermediate lambda values ($\lambda \sim 0.4$), we find that counting for 300 droplets is a good compromise between the efficiency and precision of the lambda estimate. We use the *Verify & Sort* function of the DropMap application for the lambda count for a list of 300 randomly generated droplet indexes.
40. The concentration of analyte corresponding to the LoD is the lower limit of the quantitative range of the assay.

41. Using intracellular cell staining to visualize live cells inside the droplets is possible. This added dye facilitates the counting of cells and the determination of empty droplets for the computation of the LoD. It also verifies that the positively sorted droplets contain a cell and are not false positives. If the cells are not stained, the droplets that were classified as empty during the lambda count can be used to compute the LoD. When doing so, the number of empty droplets should not be below 150. Otherwise, it should be increased by sorting more random droplets.
42. This is our adaptation of a usual definition of the LoD [12]:

$$\mathrm{LoD} = \mathrm{mean}_{\mathrm{blank}} + 1.645 * (\mathrm{SD}_{\mathrm{blank}} + \mathrm{SD}_{\mathrm{low\ concentration\ sample}})$$

We assume that a blank and low concentration sample have the same standard deviation σ (as their SD values are very close in practice) and estimate $\mathrm{mean}_{\mathrm{blank}}$ and σ using the population of empty droplets at the first time point in each measurement. The relocation signals from the empty droplet population are normally distributed, but outliers (due to fluorescent clumps for example) whose signals are not part of this normal distribution might be present. We use $\mathrm{median}_{\mathrm{empty\ droplets}}$ and $k^{*}\mathrm{MAD}_{\mathrm{empty\ droplets}}$ to estimate $\mathrm{mean}_{\mathrm{blank}}$ and σ, respectively, reliably despite the presence of those potential outlier values [13, 14]. The computing of an LoD for each individual measurement ensures that it is adapted to the exact conditions of the measurement since the assay might vary slightly from measurement to measurement due to deviations in pipetting or mixing, secretion of the analyte before encapsulation, or slightly different focal planes.
43. The coefficient 1.645 determines the level of precision and is arbitrarily chosen. In a normally distributed population of empty droplets, this corresponds to having only 0.05% of droplets with a signal above the LoD. Depending on the frequency of secreting cells in the analyzed population, it is possible to change this coefficient (lowering it when sorting for common events and increasing it for very rare events). Typically, we analyze around 40,000 droplets in a measurement so that the number of true false positives would be around 20 per measurement. This number is further reduced by adding the other two criteria in Subheading 3.6, **step 3**.
44. These criteria are suited for the IgG and IFN-γ secretion assays and might need to be adapted when changing the analyte. In particular, the noise criterium is quite stringent and could be reduced for an analyte where the calibration curve has a lower initial slope. Additionally, the quantitative range of the assay should match the analyte concentrations commonly reached by secreting cells in the measurement, otherwise, these criteria are not relevant.

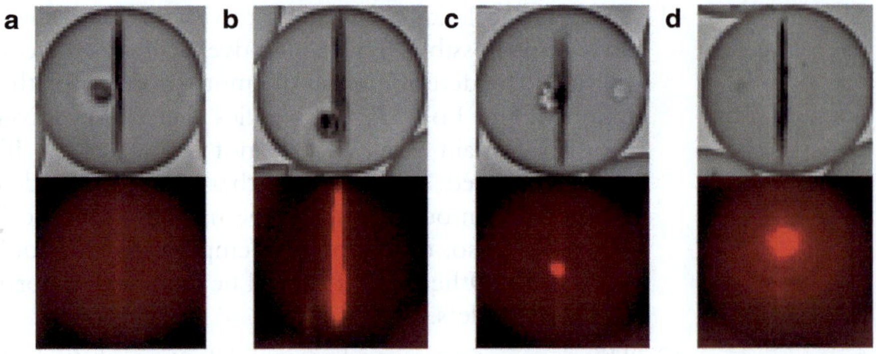

Fig. 5 Images of single droplets from an IgG secretion measurement of a murine B cell sample, after 32 min; top: brightfield channel, bottom: red fluorescence channel. (**a**) Droplet containing a cell that does not secrete IgG. (**b**) Droplet containing a cell that secretes IgG (fluorescence relocation on the beadline) should be positively evaluated during manual verification. (**c**) Droplet containing a cell with relocation of the fluorescent detection reagent on the cell should be sorted out. (**d**) Droplet containing a red fluorescent clump should be sorted out

45. Using the *Verify & Sort* function of the DropMap application, the positively selected droplets must be manually verified to check for the absence of anomalies (fluorescent particle, clump of multiple cells, or other anomalies). Droplets where the fluorescence relocation signal is due to fluorescence on the cell and not on the beadline should also be sorted out (*see* Fig. 5).

46. If you are using a cell staining and thus can detect empty droplets, applying Subheading 3.6, **step 3** to the empty droplets population will yield the false positive rate of the analysis.

47. This calculation assumes that the secretion rate is constant throughout the measurement. You should verify if this hypothesis applies to your measurement. If not, separate secretion rates for each time interval should be computed. For example, the secretion rates are constant for IgG-secreting cells over 32 min (*see* Fig. 4a, constant increase of the relocation signal except for the cells approaching the plateau). However, it is not the case for IFN-γ secretion over 4 h (*see* Fig. 4c, the cells seem to stop secreting after approximately 2 h, with a gradual decrease after the first hour).

48. Some droplets may contain cells secreting the analyte at a high rate, resulting in concentrations that exceed the range where the assay is quantitative (*see* Fig. 2). The time points where this happens cannot be used for precise computation of the secretion rate. Indeed, because of the variability of the signal obtained (standard deviation values in Fig. 2), when the fluorescence relocation signal approaches the plateau of the calibration curve, it is not possible to attribute a precise concentration to a given relocation value: the range of

concentrations that could correspond to this relocation is too large. To define the upper limit of the quantitative range of the assay, we set a cutoff on the fluorescence relocation equal to the plateau value from the exponential fit minus the half-width of the prediction interval (90% confidence level). The upper limit of the quantitative range is the concentration of analyte that leads to a relocation equal to the cutoff value. In the IgG secretion assay example, the fitted plateau value is 3.46, and the 90% confidence prediction bounds are ±0.61, giving a cutoff value of 2.85, corresponding to roughly 15 nM IgG. Any relocation signal above 2.85 is not used to compute the secretion rate. The first or second timepoints may already be above the cutoff value for some droplets due to a high IgG secretion. For those droplets, it is impossible to compute a precise secretion rate, so they are classified as containing a high-secreting cell (secreting more than 500 IgG/s in this example, *see* Fig. 4a, b, qualitative range). If a large proportion of the secreting cells are high secretors, or if a precise measurement of their secretion rate is necessary, the quantitative range of the assay should be modified (*see* **Notes 8** and **11**).

Acknowledgments

This work was supported by the European Research Council (grant no. 803363) and a Swiss National Foundation project grant (grant no. 310030_197619). This work has benefited from the technical contribution of the joint service unit CNRS UAR 3750.

References

1. Saletti G, Çuburu N, Yang JS, Dey A, Czerkinsky C (2013) Enzyme-linked immunospot assays for direct ex vivo measurement of vaccine-induced human humoral immune responses in blood. Nat Protoc 8:1073–1087
2. Jung T, Schauer U, Heusser C, Neumann C, Rieger C (1993) Detection of intracellular cytokines by flow cytometry. J Immunol Methods 159:197–207
3. Manz R, Assenmacher M, Pfluger E, Miltenyi S, Radbruch A (1995) Analysis and sorting of live cells according to secreted molecules, relocated to a cell-surface affinity matrix. Proc Natl Acad Sci USA 92:1921–1925
4. Han Q, Bradshaw EM, Nilsson B, Hafler DA, Love JC (2010) Multidimensional analysis of the frequencies and rates of cytokine secretion from single cells by quantitative microengraving. Lab Chip 10:1391–1400
5. Ma C, Fan R, Ahmad H, Shi Q, Comin-Anduix B, Chodon T et al (2011) A clinical microchip for evaluation of single immune cells reveals high functional heterogeneity in phenotypically similar T cells. Nat Med 17:738–743
6. Mazutis L, Gilbert J, Ung WL, Weitz DA, Griffiths AD, Heyman JA (2013) Single-cell analysis and sorting using droplet-based microfluidics. Nat Protoc 8:870–891
7. Parisi G, Saco JD, Salazar FB et al (2020) Persistence of adoptively transferred T cells with a kinetically engineered IL-2 receptor agonist. Nat Commun 11:660
8. Gerard A, Woolfe A, Mottet G, Reichen M, Castrillon C, Menrath V et al (2020) High-throughput single-cell activity-based screening and sequencing of antibodies using droplet microfluidics. Nat Biotechnol 38:715–721

9. Bucheli OTM, Sigvaldadóttir I, Eyer K (2021) Measuring single-cell protein secretion in immunology: technologies, advances, and applications. Eur J Immunol 51(6): 1334–1347
10. Bounab Y, Eyer K et al (2020) Dynamic single-cell phenotyping of immune cells using the microfluidic platform DropMap. Nat Protoc 15:2920–2955
11. Eyer K et al (2017) Single-cell deep phenotyping of IgG-secreting cells for high-resolution immune monitoring. Nat Biotech 35(10): 977–982
12. Armbruster DA, Pry T (2008) Limit of blank, limit of detection and limit of quantitation. Clin Biochem Rev 29(Suppl 1):S49–S52
13. Rousseeuw P, Croux C (1993) Alternatives to the median absolute deviation. J Am Stat Assoc 88(424):1273–1283
14. Leys C et al (2013) Detecting outliers: do not use standard deviation around the mean, use absolute deviation around the median. J Exp Soc Psych 49(4):764–766

Chapter 10

Magnetic Microtweezers for High-Throughput Bioseparation in Sub-Nanoliter Droplets

Simon Dumas, Lucile Alexandre, Mathilde Richerd, Marco Serra, and Stéphanie Descroix

Abstract

Multiomics studies at single-cell level require small volume manipulation, high throughput analysis, and multiplexed detection, characteristics that droplet microfluidics can tackle. However, the initial step of molecule bioseparation remains challenging. Here, we describe a unique magnetic device to trap and extract magnetic particles in sub-nanoliter droplets, for compartmentalisation of detection steps. Relying on electrodeposition of NiFe structures and microfluidic manipulation, the extraction of 1 μm diameter magnetic particles was achieved at high throughput (20 droplets per second) with an efficiency close to 100% in 450 pL droplets. The first demonstration of its adaptability to single-cell analysis is demonstrated with the extraction of mRNA. Using a purified nucleic acid solution, this unique magnetic configuration was able to reach a RNA extraction rate of 72%. This is the first demonstration of a physical separation in droplets at high throughput at single-cell scale.

Key words Droplet microfluidics, Magnetic micro-tweezers, RNA purification

1 Introduction

Miniaturization allows playing with the kinetics of reactions, reducing time of analysis, volumes of sample and reagents, and risks of contamination coupled with increasing the sensitivity of detection. Development of miniaturized systems has supported this shift, reinforced by a better comprehension of the physics at micro-scale [1]. Microfluidics often finds applications in diagnostics as it offers the possibility to work with small quantities of reagents and samples at limited costs [2]. The fabrication of micro-tools enables the implementation of various steps of laboratory work in microfluidics systems, and translation of fastidious protocols to a single instrument.

Working with a volume range from microliters to picoliters can be coupled with compartmentalization of reagent via droplet-based

Valérie Taly et al. (eds.), *Microfluidics Diagnostics: Methods and Protocols*, Methods in Molecular Biology, vol. 2804, https://doi.org/10.1007/978-1-0716-3850-7_10,

microfluidics [3, 4]. As with many microfluidic approaches, droplet encapsulation allows the manipulation of small samples in a confined environment, reducing time of analysis and contamination risks, opening the door to scale-up, parallelization, and multiplexing. Droplet microfluidics has progressed toward high-impact studies with complex applications [5] at high throughput such as biomarkers detection [6], drug screening [7], 3D cell culture [8, 9], or single-cell measurements [10, 11]. Generation, merging, mixing, sorting, and splitting of droplets can be achieved at micro-scale, relying on micro-fabrication and surface treatment development [12]. However, some operations remain challenging to achieve such as extraction and purification steps, as they often rely on a physical separation.

To tackle this challenge, the magnetic tweezers technology has been developed by our team for the extraction and redispersion of magnetic beads inside droplets [13]. The versatility of a magnetic solid-phase extraction makes it an obvious candidate for translation to micro-scale. The technology of magnetic tweezers relies on the control by tunable electric coil of an external magnetic field focused on the path of droplets. Magnetic particles inside the droplet are then confined in a compact cluster that can be extracted from the initial droplet, washed, and released in a second droplet by switching off the magnetic field. This process has already shown its efficiency for biomarker extraction and detection [14–16].

Here, we present a new generation of magnetic tweezers for the manipulation of magnetic micro-particles in sub-nanoliter droplets and its application to purification of small quantities of mRNA compatible with single-cell analysis. This new approach is based on NiFe micro-tweezers, which are fabricated by photolithography and electrodeposition and are actuated by an external permanent magnet. Modification of the geometry and structure of the micro-tweezers tackle the challenge to work at high throughput with sub-nanoliter droplets without sacrificing the sensitivity. This new design has allowed us to work at single-cell level, with 0.5 nL droplet at a throughput of 20 Hz [17]. This unique magnetic configuration has bridged the gap between high throughput micro-fluidic sample manipulation and multiomics analysis at single-cell level.

2 Materials

2.1 Substrate Preparation Before Plating

1. Large microscope glass slides 50 × 75 mm (Corning).
2. Sputtering machine equipped with Cu and Ti targets.
3. Photolithography equipment (chromium mask, mask aligner, hotplates, spin coater).

4. Photoresist adhesion promoter (TI-Prime, Microchemicals).
5. Negative photoresist (AZ 125 nXT, Microchemicals) and developer (AZ 726 MIF, Microchemicals).

2.2 Electroplating

1. Plating solution: dissolve in 700 mL of water the following reagents: 250 g $NiSO_4 \cdot 6H_2O$, 5 g $FeSO_4 \cdot 7H_2O$, 25 g Boric acid, 2 g Saccharin, 0.1 g Sodium dodecyl sulfate. Place the beaker on a hotplate at 40 °C while stirring until complete dissolution, and fill with water up to 1 L. Filter the solution through Stericup (Millipore). The solution must be dark green and can be kept and reused for at least 3 months at room temperature.
2. DC generator (ALR3002M, ELC).
3. Conductive copper tape (1181/9.5, 3 M).
4. Pure nickel anode, 100 × 60 × 2 mm (Goodfellow).
5. UV Ozone cleaner (Jelight).
6. Arduino Uno.
7. Relay board (Radiospares, 843–0834).
8. Photoresist stripping solution, Technistrip P1316 (Technic).

2.3 Microfluidic Chip Preparation

1. Photolithography equipment (chromium mask, mask aligner, hotplates, spin coater, SU-8, developer).
2. PDMS and curing agent (Sylgard 184).
3. Mask aligner (MJB4, Süss) or custom tabletop aligner [18, 19].
4. Tetraethyl orthosilicate (TEOS).
5. 1H,1H,2H,2H-Perfluorooctyl Trichlorosilane.

2.4 Microfluidic Device Operation

1. Syringe pumps (Cetoni).
2. Bright-field inverted microscope.
3. Biopsy punches (diameter 0.75, 6, and 8 mm).
4. 1 mL syringes (Omnifix, Braun) and beveled needles 23G and 27G (Terumo).
5. PTFE tubings (Adtech BIOBLOCK/05 and BIOBLOCK/13).
6. Neodymium magnets (5 mm cube, N50, and 10 mm cube, N42, Supermagnete).
7. Carrier oil: HFE-7500 with 2% fluorosurfactant, Emulseo.

2.5 mRNA Extraction in Droplets

1. Streptavidin Dynabeads MyOne C1 (Thermofisher, 65,001).
2. Dynabeads solution A: 0.1 M NaOH, 50 mM NaCl.
3. Dynabeads solution B: 100 mM NaCl.

4. 2X Binding/Wash buffer (10 mM TrisHCl pH 7.5, 1 mM EDTA, 2 M NaCl, 0.05% *v/v* Triton X-100).
5. 1X Binding/Wash buffer (Dilute 2X buffer in same volume of water).
6. Biotinylated oligo-dT30VN (5′-/5BiotinTEG/AAGCAGTG GTATCAACGCAGAGTACTTTTTTTTTTTTTTTTTTTT TTTTTTTTTVN).
7. Cell Resuspension Buffer: 15% *v/v* Optiprep, 0.75% BSA in PBS.
8. 1.5X Lysis/Binding Buffer: 150 mM TrisHCl pH 7.5, 750 mM NaCl, 15 mM EDTA, 0.075% *w/v* Triton X-100, 0.75% *w/v* BSA, 1 U/μL Ribolock RNAse Inhibitor. Prepare fresh and keep on ice until use.
9. Emulsion breaking solution: 20% *v/v* 1H,1H,2H,2H-perfluoro-1-octanol in HFE-7500.

2.6 mRNA Quantification

1. qPCR equipment (Quantstudio 3, Applied Biosystems).
2. RT-qPCR kit (CellsDirect One-Step qRT-PCR, Thermofisher, 11,753,100).
3. ACTB qPCR primers (Fw 5′- GGATGCAGAAGGAGAT CACTG, Rv 5′-CGATCCACACGGAGTACTTG).
4. ACTB TaqMan fluorescent probe (5′−/56-FAM/AAGAT-CAAG/ZEN/ATCATTGCTCC/3IABkFQ/).

3 Methods

The microfabrication of substrates and chips requires a clean room environment and special equipment to perform photolithography and physical vapor deposition (PVD). This includes a chromium mask, a mask aligner, a spin coater, hotplates, a sputtering machine, a plasma chamber, and a vacuum desiccator. First, the magnetic micro patterns are fabricated using photolithography and electroplating on glass slides. Second, the microfluidic chip is obtained by PDMS molding on a SU-8 patterned wafer and is aligned and bonded onto the magnetic patterns (Fig. 1).

Design considerations: The designs of the chip and the patterns must be complementary, and a chamber must be provided to accommodate the magnets. For the pattern design, an area for connecting the alligator clip during electroplating must be anticipated. This area must remain dry during electroplating and is preferably placed at least 1 cm away from the patterns.

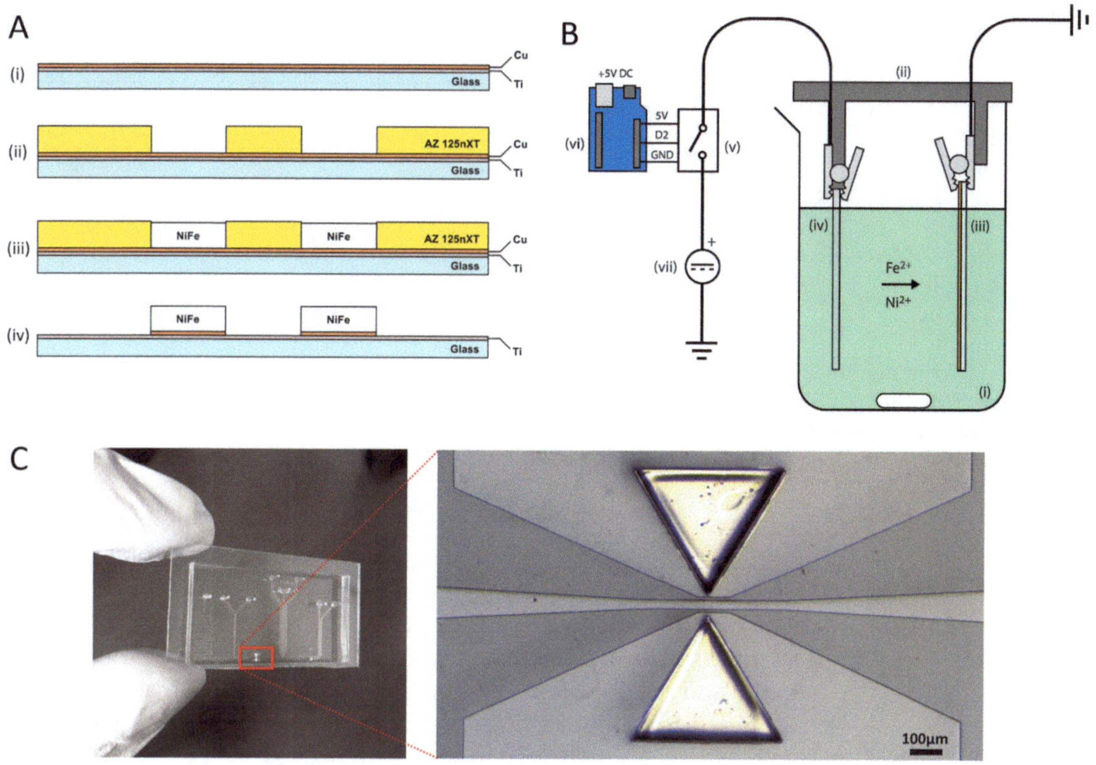

Fig. 1 (**a**) Fabrication of the micromagnet substrate (i) Sputtering of a Ti:Cu layer on glass slide (ii) Photolithography (iii) Electroplating (iv) Stripping of photoresist and Cu, and TEOS treatment. (**b**) Electroplating process (i) Beaker filled with the plating solution (ii) 3D-printed holder (iii) Substrate (iv) Nickel anode (v) Relay used as a programmable switch (vi) Arduino board (vii) Stabilized DC source (**c**) Final chip after alignment and bonding of the PDMS channels on the micromagnet substrate

3.1 Preparation of the Substrates for Electroplating

1. Clean a glass slide with acetone, isopropanol, and proceed to seed layer deposition by sputtering of 20 nm Ti and 200 nm Cu (*see* **Note 1**).
2. Apply TI-Prime by spin coating at 4000 rpm for 30 s and dry the substrate on hotplate at 95 °C for 1 min.
3. Apply negative photoresist AZ 125 nXT by spin coating (*see* **Note 2**) and bake for 4 min at 95 °C and 10 min at 125 °C. Let cool down to room temperature.
4. Using a mask aligner, expose the substrate to 2 J/cm^2 of UV-light (i-line, 365 nm) through a chromium mask. Develop the photoresist by immersion in AZ 726 MIF for 2 min, rinse with water, and dry. The substrate can be kept protected from light for at least 3 months.

3.2 Electroplating of NiFe Structures

This step may require optimization and various problems can be encountered. Several parameters can significantly influence the result in terms of deposit quality, magnetic properties, or thickness.

Table 1
Arduino program loop

```
int digout = 2; // Use digital output 2 on Arduino board
int counter = 0;
void setup() {
  pinMode(digout, OUTPUT);
}
void loop() { // Every 5 min, make a 15 s break. Every hour, make a 2 min break.
  counter = 0;
  while(counter != 12) {
    compteur ++;
    digitalWrite(digout, LOW);
    delay(300000); // 5 min
    digitalWrite(digout, HIGH);
    delay(15000); // 15 s
  }
  delay(105000); // 2 min
}
```

They include deposition time, electrical current, pattern design, stirring speed, bath temperature and pH, presence of dust in the bath, and evaporation. The protocol was adapted from [20] and modified for higher thicknesses. It was observed that the deposit will detach after long periods of deposition. For this reason, a programmable switch was added to periodically open and close the circuit, decreasing the internal stress within the deposit.

1. Place a beaker containing around 700 mL of plating solution on a hotplate set at 40 °C with continuous stirring.
2. Conduct UV-Ozone treatment for 10 min on the previously developed substrate to improve pattern wettability, and wet them using a wash bottle filled by 1% SDS.
3. Apply copper tape on the substrate (using a dedicated dry area) and on the nickel anode, and connect them with alligator clips.
4. Connect them to the DC generator, the substrate to minus, and anode to plus, with the relay used as an interrupter to programmatically open and close the circuit. Connect the relay to the Arduino: GND pin to GND, power pin to +5 V, and the control pin to digital output 2.
5. Immerse the substrate and the anode in the bath, they must face each other with a distance of 5–7 cm, using a 3D printer holder. Immediately turn the generator on at minimum power. Increase the current to reach a current density of 7 mA/cm^2 (*see* **Notes 3** and **4**).
6. Fill the beaker with a plating solution until all the patterns are immersed. Ensure that the copper tape and the alligator clip remain dry.

7. Power the Arduino board up to launch the program loop described in Table 1.
8. Let the substrate in the plating solution with continuous stirring (*see* **Note 5**) until the thickness reaches 35 μm. The expected deposition time is 12 h, with a rate of 3 μm/h (*see* **Notes 6** and 7).
9. Clean the substrate with water and immerse in TechniStrip P1316 at 70 °C while stirring until the photoresist and copper dissolve completely. This step can be repeated after 20 min by renewing the solvent to remove eventual residues (*see* **Note 8**).

3.3 Preparation of the Microfluidic Chips

The fabrication of the PDMS counterpart is performed by conventional soft lithography, using 4″ wafers and SU-82035 to achieve a thickness of 35 μm. PDMS with a curing agent ratio of 1:10 is used. This method is not detailed here as it is extensively described in the microfluidics literature.

1. Prepare the PDMS chips by soft lithography. To facilitate a precise alignment with the magnetic patterns, the PDMS slabs must be as flat as possible. This can be achieved either by leveling the oven during PDMS curing or by curing the PDMS between two parallel plastic plates [21].
2. A thin titanium layer remains on the magnetic substrate, which inhibits plasma bonding with the PDMS, and the magnetic substrate needs to be treated with a silane. Activate the magnetic substrate with O_2 plasma and place it under vacuum in the presence of tetraethyl orthosilicate (TEOS) for 30 min.
3. Activate PDMS and magnetic substrate with O_2 plasma before alignment.
4. Align and bond the two parts together using either a mask aligner [21] or a custom tabletop aligner [18, 19].
5. Press gently on the PDMS and place the bonded chip in an oven (or hotplate) at 70 °C for at least 10 min to ensure complete bonding.
6. The channels must be silanized to render them hydrophobic: prepare and filter a solution of 1% *v/v* 1H,1H,2H,2H-Perfluorooctyl trichlorosilane in HFE-7500 and flow it in the microfluidic channels using a syringe. After no more than 5 min, rinse with pure HFE-7500, blow air, and place back the chips at 70 °C to evaporate the remaining oil. Chips can be stored for months at room temperature, protected from dust.

3.4 Preparation of Collection/Reinjection Reservoirs

1. Using a 8 mm biopsy punch, cut a cylinder from a 5–10 mm thick PDMS slab.
2. Punch two holes using a 0.75 mm biopsy punch, clean with isopropanol.

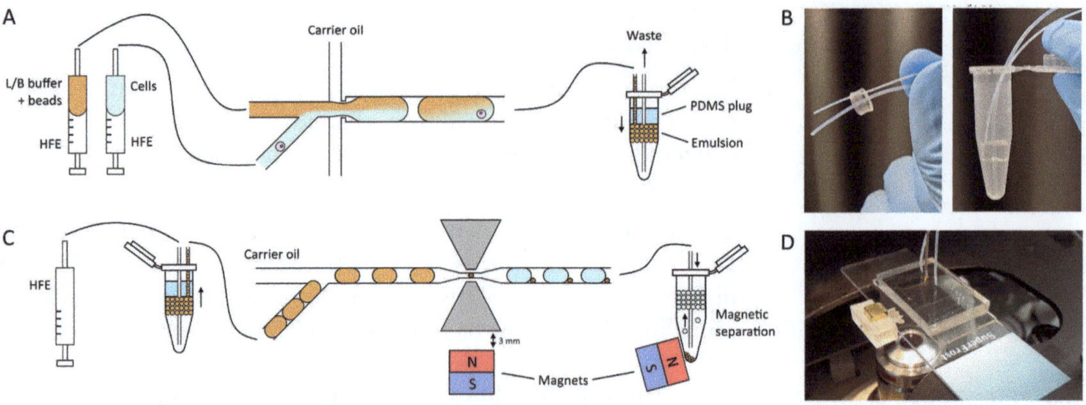

Fig. 2 Microfluidic chip operation (**a**) Single-cell encapsulation with beads and lysis buffer. (**b**) Preparation of the collection/reinjection reservoir. (**c**) Reinjection of the droplets in the magnetic microtweezer chip. (**d**) Permanent magnet positioning using a custom 3D-printed holder

3. Using a tweezer, pass a 0.56 mm (I.D.) tubing through one hole and a 0.3 mm (I.D.) tubing through the other.
4. Using a tweezer, push the PDMS plug in a 1.5 mL microcentrifuge tube.
5. Using a tweezer, push or pull the tubings so that the wider one reaches the bottom of the tube, and the thinner one end is at the PDMS interface (Fig. 2b).
6. Prepare a syringe filled with carrier oil with a 23G needle and connect it to the wider tubing using a tweezer. Push the plunger to fill the reservoir until the oil drops from the thinner tubing. Keep the syringe connected to the reservoir.

3.5 Preparation of Oligo-dT Conjugated Beads

The extraction protocol is optimized for Dynabeads MyOne. These beads are not commercialized with oligo-dT functionalization and need to be conjugated in another way.

1. Take 80 μL of Dynabeads MyOne Streptavidin C1 (80 μg of beads).
2. Using a magnet, wash two times in 1 mL of Dynabeads solution A.
3. Perform two additional washing with Dynabeads solution B.
4. Discard supernatant and add 80 μL of 2X Binding/Wash buffer.
5. Add 80 μL of 100 μM biotinylated oligo-dT30VN.
6. Incubate 20 min at room temperature with rotation.
7. Wash three times with 1X Binding/Wash buffer.

3.6 Droplet Generation and Cell Encapsulation

For cell encapsulation, a drop maker with two aqueous inlets is required. In this part, droplets of 0.5–1 nL are generated and stored in the reservoir. The droplets contain cells, a lysis/binding buffer, and oligo-dT magnetic beads (Fig. 2a).

1. Harvest cells using a conventional cell culture protocol (not detailed here).
2. Resuspend the cells in 70 μL of Cell Resuspension Buffer at a concentration of 600 cells/μL.
3. Wash the oligo-dT conjugated beads with 1.5X Lysis/Binding Buffer and resuspend them in 200 μL of the same buffer at a concentration of 60 μg/μL (*see* **Note 9**).
4. Fill two syringes halfway with HFE-7500 and with the above-mentioned solutions. Mount the two syringes on the syringe pump vertically to allow the aqueous phase to remain above (*see* **Note 10**). Mount a third syringe filled with carrier oil on the syringe pump.
5. Place the encapsulation microfluidic chip under a microscope. Plug 27G needles on the syringes and connect them to the chip with 0.3 mm (I.D.) tubings.
6. Generate 0.5 nL droplets by using the following flow rates: 200 μL/h for the buffer with beads, 100 μL/h for the cells, and 200 μL/h for the carrier oil (*see* **Note 11**).
7. When the generation is stable, connect the collection/reinjection reservoir to the chip and disconnect the syringe at the reservoir outlet. The emulsion should fill the reservoir from the top. Keep the reservoir on ice or in a cold block during the whole experiment.
8. When the generation is finished, disconnect the reservoir from the chip and connect its outlet back to the syringe. Keep the reservoir on ice.

3.7 Magnetic Beads Extraction Through the Magnetic Microtweezers

In this part, the emulsion is reinjected into the chip with the micro tweezers. The droplets are spaced with oil and the beads are extracted by the tweezers. After the extraction, the droplets and the extracted bead clusters are collected and separated in an external tube equipped with a permanent magnet (Fig. 2c).

1. Place the extraction microfluidic chip under a microscope with the 5 mm cubic magnet facing the tweezers at a distance of 3 mm using a 3D-printed holder (Fig. 2d).
2. Mount the reservoir syringe to the syringe pump, and connect the reservoir to the chip.
3. Inject the emulsion and spacing oil at 40 μL/h and 20 μL/h, respectively. The droplet reinjection frequency should be around 40 droplets/s. The extraction of magnetic beads takes

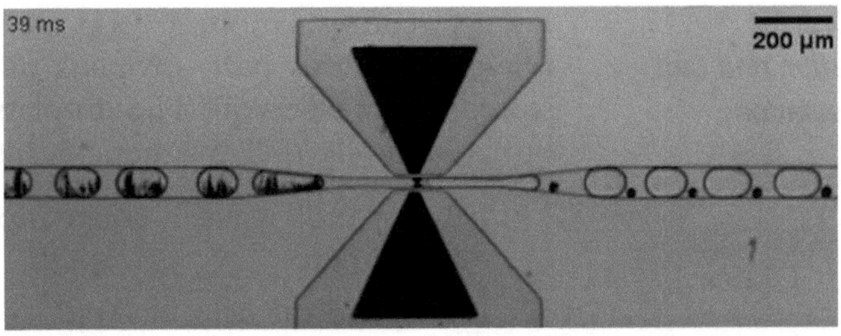

Fig. 3 Extraction of magnetic beads from the droplets. The droplets containing beads arrive from the left. The beads are extracted as a compact cluster, which is then pushed by the following droplet

place at the micro tweezers. Figure 3 shows what a good extraction looks like.

4. When the extraction is stable, connect the outlet tubing to a 1.5 μL microcentrifuge tube filled with 300 μL carrier oil. Place the 10 mm square neodymium magnet close to the tube bottom to separate the bead cluster from the droplets (*see* **Note 12**). Keep the collection tube on ice during extraction.

3.8 Assessment of mRNA Purification Rate by RT-qPCR

This part aims at assessing the efficiency of mRNA purification of the above protocol. The quantity of mRNA in the droplets is evaluated by RT-qPCR on the emulsion before and after extraction through the micro tweezers.

1. Using a micropipette, transfer 20 μL of the outlet emulsion in a new tube.
2. Disconnect the reinjection reservoir from the chip and dispense 20 μL of the inlet emulsion in another tube. Ensure that the two tubes contain the same volume of emulsion.
3. Add in each tube 50 μL of emulsion breaking solution. Vortex and centrifuge briefly.
4. Using a micropipette, collect 15 μL of the aqueous phase and transfer in another tube (*see* **Note 13**). Keep the tubes on ice.
5. Prepare a serial dilution (1:10, 1:100, 1:1000) of an mRNA sample of known concentration, which will be used as a ladder to convert the qPCR Ct values to mRNA concentration.
6. Prepare the RT-qPCR master mix using the RT-qPCR kit, as follows (*see* **Note 14**):
   - 50 μL Reaction mix (2X)
   - 2 μL Superscript Platinium Taq mix
   - 2 μL Forward primer (10 μM)
   - 2 μL Reverse primer (10 μM)
   - 2 μL TaqMan probe (10 μM)

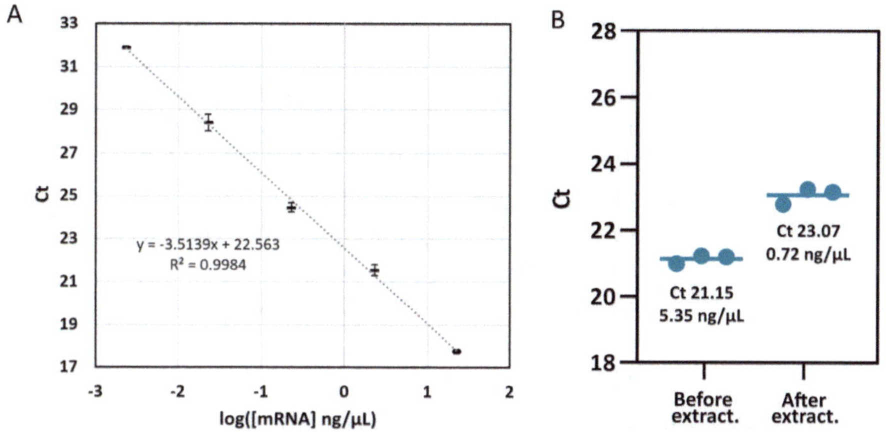

Fig. 4 Results and analysis of RT-qPCR. (**a**) Calibration curve was obtained with the ladder as a sample. (**b**) Results for the droplet sample, before and after the extraction through the micro tweezers

- 0.05 μL ROX dye
- 37.8 μL Nuclease free water

7. Perform RT-qPCR in reactions containing 24 μL of the above master mix with 1 μL on undiluted droplet content from **step 4** (or the ladder dilution). Use the following program:
   - 50 °C 15 min
   - 95 °C 2 min
   - 95 °C 15 s, 60 °C 30 s | × 45
8. Use a linear interpolation on the ladder Cts in log scale to get the conversion relation between the Ct and the mRNA concentration (Fig. 4a).
9. Using this conversion, calculate the mRNA quantity in the emulsions before and after extraction to evaluate the purification efficiency (Fig. 4b).

4 Notes

1. The deposition protocol will depend on the available equipment. We use 75 W, 15 min for Ti and 100 W, 5 min for Cu at room temperature. Some variations of seed layer thickness over experiments can be observed and are not critical. The back and edges of the substrate must be covered with tape to render only one side conductive.
2. The spin coating protocol must be optimized for each thickness. We obtained a thickness of around 50 μm with the following protocol: 300 rpm 5 s, 2600 rpm 2 s, and 2800 rpm 8 s. The baking time also needs to be optimized: after cooling

down, the photoresist needs to be slightly tacky but should not stick on gloves.

3. The generator must be turned on as soon as possible after immersion of the substrate in the bath to avoid copper dissolution in the acidic bath.
4. The current density is calculated with the area of patterns. The area has to be large enough (>0.1 cm^2) to be in the operating range of the DC generator.
5. Stirring must be strong enough to continuously renew the solution within the patterns. Use the highest stirring speed which does not induce bubbles, foaming, or instabilities.
6. The deposition may vary with the design, stirring speed, electrode distance, and bath age. A control measurement can be done after a few hours of deposition using a stylus profilometer to check the deposition rate. However, care must be taken to keep the substrate wet as much as possible as the photoresist might crack within minutes after being removed from water.
7. After a long deposition step, the water in the plating bath may have evaporated. Fill the plating bath with water back to its original level, and keep it in a bottle at room temperature.
8. The magnetic substrate can be cut into several parts if used to make several microfluidic chips. Use a laser cutter or a tungsten tip. The cut can be done before photoresist stripping to keep the substrate clean.
9. The concentrations are calculated for a final concentration in droplets of 40 μg/μL of beads and 200 cells/μL. The flow ratio between the bead and droplet solution is 2:1. This way, a cell is encapsulated for every 10 droplets of 0.5 nL.
10. When filling the syringe with the cell solution, do not use a thin needle that will cause cell lysis by shearing.
11. Beads will sediment in the tubings. The generation has to be done quickly.
12. The magnet can be locked to the tube using glue, tape, or a custom 3D-printed holder. A commercial magnetic rack might be used but some are often not powerful enough.
13. In the bead-containing tube, redisperse the beads homogeneously in the aqueous phase by gently pipetting up and down without disturbing the oil/water interface.
14. The master mix quantity: prepare as much master mix as needed, depending on the number of samples and ladder. Each reaction is preferably done in triplicates.

Acknowledgments

The authors acknowledge all members of the IPGG technological platform for their help with microfabrication, especially Olivier Lesage for his support in photolithography and sputtering and Bertrand Cinquin for EDX spectrometry. They also thank Koceila Aizel for support with electroplating, and Fanny Tabarin and Aude Battistella for training them on cell and molecular biology (Institut Curie, UMR168), David Hrabovsky (MPBT platform, Sorbonne University) for magnetometry measurements, and Olivier Lefebvre (C2N, Univ. Paris-Saclay) for helpful advice on electroplating. This work was supported by the Institut Pierre-Gilles de Gennes (équipement d'excellence and LABEX, "Investissement d'avenir," program ANR-10-EQPX-34). This project has received funding from the European Union's Horizon 2020 research and innovation program under the Marie Skodowska-Curie grant agreement No 896313.

References

1. Tabeling P (2005) Introduction to microfluidics. OUP, Oxford
2. Whitesides GM (2006) The origins and the future of microfluidics. Nature 442:7101
3. Atencia J, Beebe DJ (2005) Controlled microfluidic interfaces. Nature 437:7059
4. Stone HA, Stroock AD, Ajdari A (2004) Engineering flows in small devices: microfluidics toward a lab-on-a-chip. Annu Rev Fluid Mech 36:381–411
5. Ding Y, Howes PD, deMello AJ (2020) Recent advances in droplet microfluidic. Anal Chem 92:132–149
6. Fitarelli-Kiehl M et al (2018) Denaturation-enhanced droplet digital PCR for liquid biopsies. Clin Chem 64:1762–1771
7. Seeto WJ, Tian Y, Pradhan S, Minond D, Lipke EA (2022) Droplet microfluidics-based fabrication of monodisperse poly(ethylene glycol)-fibrinogen breast cancer microspheres for automated drug screening applications. ACS Biomater Sci Eng 8:3831–3841
8. Tomasi RF-X, Sart S, Champetier T, Baroud CN (2020) ndividual control and quantification of 3D spheroids in a high-density microfluidic droplet array. Cell Rep 31:107670
9. Langer K, Joensson HN (2020) Rapid production and recovery of cell spheroids by automated droplet microfluidics. SLAS Technol 25:111–122
10. Kang HM et al (2018) Multiplexed droplet single-cell RNA-sequencing using natural genetic variation. Nat Biotechnol 36:1
11. Gierahn TM et al (2017) Seq-Well: portable, low-cost RNA sequencing of single cells at high throughput. Nat Methods 14:395–398
12. Teh S-Y, Lin R, Hung L-H, Lee AP (2008) Droplet microfluidics. Lab Chip 8:198–220
13. Ali-Cherif A, Begolo S, Descroix S, Viovy J-L, Malaquin L (2012) Programmable magnetic tweezers and droplet microfluidic device for high-throughput nanoliter multi-step assays. Angewandte Chemie 124:10923–10927
14. Ferraro D et al (2016) Microfluidic platform combining droplets and magnetic tweezers: application to HER2 expression in cancer diagnosis. Sci Rep 6:1
15. Mai TD et al (2018) Single-step immunoassays and microfluidic droplet operation: Towards a versatile approach for detection of amyloid-β peptide-based biomarkers of Alzheimer's disease. Sens Actuators B Chem 255:2126–2135
16. Hajji I et al (2020) Droplet microfluidic platform for fast and continuous-flow RT-qPCR analysis devoted to cancer diagnosis application. Sens Actuators B Chem 303:27171
17. Dumas S, Richerd M, Serra M, Descroix S (2022) Magnetic microtweezers: a tool for high-throughput bioseparation in sub-nanoliter droplets. Adv Mater Technol 2200747

18. Li X et al (2015) Desktop aligner for fabrication of multilayer microfluidic devices. Rev Sci Instrum 86:075008
19. Guglielmotti V, Saffioti NA, Tohmé AL, Gambarotta M, Corthey G, Pallarola D (2022) A portable and affordable aligner for the assembly of microfluidic devices. HardwareX 12:e00348
20. Tseng P, Judy JW, Di Carlo D (2012) Magnetic nanoparticle-mediated massively-parallel mechanical modulation of single-cell behavior. Nat Methods 9:1113–1119
21. Cottet J, Vaillier C, Buret F, Frénéa-Robin M, Renaud P (2017) A reproducible method for μm precision alignment of PDMS microchannels with on-chip electrodes using a mask aligner. Biomicrofluidics 11:064111

Part V

Microfluidic Chips for Analysis of Bacteria and Virus

Chapter 11

A Thermoplastic Microsystem to Perform Antibiotic Susceptibility Testing by Monitoring Oxygen Consumption

Petra Jusková, André Kling, Steven Schmitt, and Petra S. Dittrich

Abstract

Antibiotic susceptibility testing (AST) is a routine procedure in diagnostic laboratories to determine pathogen resistance profiles toward antibiotics. The need for fast and accurate resistance results is rapidly increasing with a global rise in pathogen antibiotic resistance over the past years. Microfluidic technologies can enable AST with lower volumes, lower cell numbers, and a reduction in the sample-to-result time compared to state-of-the-art systems. We present a protocol to perform AST on a miniaturized nanoliter chamber array platform. The chambers are filled with antibiotic compounds and oxygen-sensing nanoprobes that serve as a viability indicator. The growth of bacterial cells in the presence of different concentrations of antibiotics is monitored; living cells consume oxygen, which can be observed as an increase of a luminesce signal within the growth chambers. Here, we demonstrate the technique using a quality control *Escherichia coli* strain, ATCC 35218. The AST requires 20 μL of a diluted bacterial suspension ($OD_{600} = 0.02$) and provides resistance profiles about 2–3 h after the inoculation. The microfluidic method can be adapted to other aerobic pathogens and is of particular interest for slow-growing strains.

Key words Antibiotic susceptibility testing, Microfluidics, Thermoplastic microfluidic devices

1 Introduction

In recent years, the increased occurrence of resistant or multiresistant bacterial strains has led to several infectious diseases becoming difficult to treat or close to untreatable [1, 2]. To address the increasing public health burden of antimicrobial resistance, alongside the development of new antibiotic drugs, new diagnostic approaches are urgently needed. Currently, the clinical labs largely rely on growth-based assays such as broth and agar microdilution, disk diffusion technique, exponential gradient E-test, and automatized systems, such as Vitek 2 [3]. Typically, growth-based techniques require long incubation times in order to obtain the required sample biomass, which could lead to a critical delay in the resistance information necessary for patient treatment. Moreover, as many genes associated with resistance are well-characterized, microbial

Valérie Taly et al. (eds.), *Microfluidics Diagnostics: Methods and Protocols*, Methods in Molecular Biology, vol. 2804,
https://doi.org/10.1007/978-1-0716-3850-7_11,

resistance can be determined by polymerase chain reaction and related techniques with high speed and sensitivity. However, these techniques are relatively expensive, require well-trained personnel, and cannot be applied generally to all resistance mechanisms [4, 5].

The miniaturization of the cultivation systems for growth-based antibiotic susceptibility testing (AST) would decrease the required cell number per assay and could dramatically shorten the overall time for the test [6, 7]. Furthermore, previous knowledge of resistance genes is not needed.

To perform AST, the following steps are needed: (1) compartmentalization of bacteria, (2) delivery of the antibiotic compound, and (3) a method to determine bacterial viability after antibiotic exposure. In particular, the first step can be conveniently realized with microfluidic technology. For example, cells can be isolated in nanoliter-volume water droplets and dispensed in an immiscible oil phase. This so-called droplet microfluidics enables high-throughput isolation of bacterial cells with single-cell resolution [8, 9]. Alternatively, cells are captured in small chambers, followed by a supply of an immiscible phase to seal the chambers [10, 11].

Individual chambers can also be sealed mechanically using the SlipChip approach [12, 13] or microfluidic valves [14, 15]. Compartmentalization in chamber systems typically does not reach the throughput of the droplet production, but the great control over the cultivation sites' position and the possibility to track individual cells using high-resolution microscopy proved advantageous in many AST systems [16, 17].

Exposure of bacterial cells to selected drugs can be performed directly at the inoculation sites [10], by dosing within the microfluidic platform [14], or in a form of a concentration gradient [17, 18].

Selecting a reliable and sensitive bacterial viability technique is a crucial part of accelerating the AST process. Due to high compatibility with microfluidic systems, assays with electrical [19, 20] or optical readouts [10, 21] are the most widely used, but other approaches, for example, cantilevers [22, 23] are explored as well.

Despite the great potential of microfluidic techniques to provide a rapid and sensitive alternative to current AST techniques, it remained difficult to translate the methods into clinical settings, because the operation of microfluidic devices is too sophisticated for routine use. In addition, the choice of the material limited the translation from "bench to bedside." The polymer polydimethylsiloxane (PDMS) is often employed for device development; however, it absorbs drugs and other hydrophobic and amphiphilic compounds, and surface properties change over time. More stable and inert materials such as glass and thermoplastics are therefore favorable. Industrially well-known fabrication methods facilitate the mass production of such microfluidic devices.

Here, we detail a protocol to perform AST using a conceptually simple microfluidic platform made in the air-tight thermoplastic material cyclic olefin copolymer (COC). We aimed at developing a platform with reduced complexity, facile operation, and with the possibility to upscale the manufacturing. The platform consists of four sets of chamber arrays, where each array is built by 144 chambers with a size of $300 \times 300 \times 75\ \mu m^3$ [24]. The first chamber array is utilized for control experiments and remains without antibiotics. The residual three chambers can be used to expose cells to MIC breakpoint concentrations (qualitative AST), different concentrations of antibiotics (quantitative AST), or different antibiotic compounds. Each chamber represents an individual cultivation site containing a growth medium and oxygen-sensing nanoprobes embedded in an agarose gel matrix. Oxygen consumption by aerobic bacterial cells growing in the presence of antibiotics is compared to the control without antibiotic exposure. The changes in the oxygen content in the medium are recorded by microscopy via luminescence of the oxygen-sensing nanoprobes. Oxygen acts as a quencher of the luminescence; thereby, oxygen consumption by viable bacteria manifests as an increase in the luminesce collected from the individual chambers (Fig. 1). We use a microscope in these studies for recording images at high resolution to observe individual cells. This is not needed for a pure AST workflow, where low-resolution imaging of the luminescence would be sufficient.

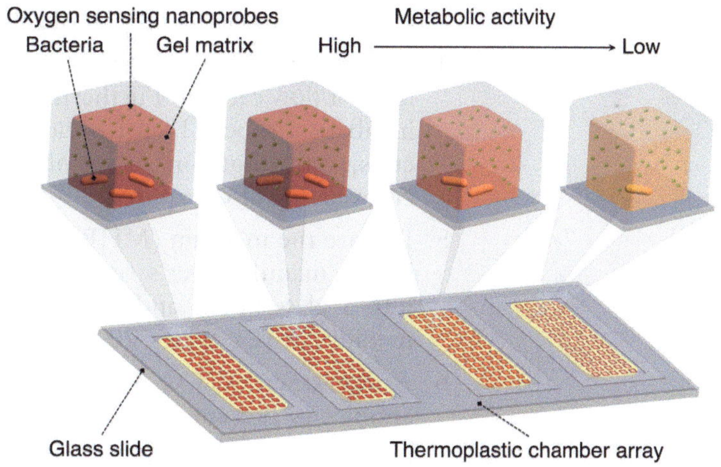

Fig. 1 Schematic representation of the platform with the chamber array for AST. The four sets of 144 cultivation chambers ($300 \times 300 \times 75\ \mu m^3$) are pre-filled with oxygen-sensing nanoprobes and antibiotics in agarose gel. The change in oxygen consumption of bacteria is determined based on luminesce of the oxygen-sensing nanoprobes

2 Materials

2.1 Bacterial Cultivation

The minimal inhibitory concentrations and clinical breakpoint concentrations for different bacteria are available in the reference guidelines [25, 26]. For the system characterization, we used one of the quality control strains *Escherichia coli* ATCC 35218. We modified the strain to overexpress the gene for the superfolder green fluorescent protein (sfGFP), for easier observation of the changes in bacterial morphology under antibiotic drug exposure. *E. coli* strain ATCC 35218 *[pSEVA271_sfgfp]* was only used for the system characterization; production of the fluorescent protein is not required for AST system functionality (*see* **Note 1**). Furthermore, this particular strain produces TEM-1 β-lactamase, which leads to resistance to β-lactam antibiotics. The use of a strain that is resistant to β-lactam antibiotics, such as amoxicillin, allows us to test for both possible situations where a pathogen is either resistant or susceptible to tested antibiotics. For the described experiments, we used a medium recommended for determining antibiotic susceptibility, i.e., Mueller Hinton Broth II, cation-adjusted medium (MHB II).

1. It is important to maintain all used materials sterile by autoclaving or by sterile filtration through the syringe filters: 0.2 μm. Plastic syringes 5–50 mL, depending on the volume of the stock solutions, are necessary for the sterile filtration process.
2. DI water for the preparation of solutions.
3. 70% ethanol in DI water or a similar disinfectant active for *E. coli* is necessary for the sterilization of materials and surfaces.
4. Bacterial incubation requires a shaking incubator and a sterile laminar flow cabinet. Furthermore, disposable materials such as incubation tubes (14 mL), inoculation loops, serological pipettes, centrifugation tubes of 10- or 50-mL volumes, and cryovials are also required.
5. Prepare the medium (MHB II) from the powder base by adding of manufacturer's recommended amount of the powder into a glass bottle and dissolved in DI water. Autoclave the medium at 121 °C for 20 min (*see* **Note 2**).
6. *E. coli* strain ATCC 35218 *[pSEVA271_sfgfp]* requires two antibiotics present in the media to maintain its resistance to β-lactam antibiotics (ampicillin sodium salt, stock solution 100 mg/mL) and to produce a green fluorescent protein (kanamycin sulfate, 10 mg/mL stock solution). Prepare both stock solutions in water and store them as 1 mL aliquots at −20 °C.
7. 60% Glycerol in DI water.

8. *Escherichia coli* strain ATCC 35218 *[pSEVA271_sfgfp]* (colonies on an agar plate).
9. Cryo-stocks: Take an agar plate containing *E. coli* strain ATCC 35218 *[pSEVA271_sfgfp]*, sterilize its outside surface, and place the plate inside the sterile laminar flow cabinet. Sterilize the surface of the bottle containing MHB II medium, 2 mL serological pipette, and 14 mL cultivation tube and place them also in the laminar flow cabinet. Using the serological pipette, transfer 2 mL of the medium to the cultivation tube and use a sterile inoculating loop to transfer five similarly looking *E. coli* colonies to this cultivation tube. Add antibiotic solutions required for the plasmid maintenance so the final concentrations would be: 50 μg/mL kanamycin sulfate and 100 μg/mL ampicillin sodium salt. Incubate the cells overnight in an incubation shaker at 37 °C and 220 rpm. The next morning, sterilize the surface of the incubation tube with cells, transfer 1200 μL of the cell suspension to a new clean tube, and add 600 μL of the 60% glycerol. Mix well and transfer 50 μL of the mixture to the cryovials. Prepare multiple aliquots so that each can be used only once per experiment without refreezing. Store the aliquots at −80 °C until use.

2.2 Antibiotic Susceptibility Testing

1. Standard laboratory equipment is necessary to perform here described AST, such as a thermomixer, a vortex mixer, and an ultrasonication bath. In this protocol, we used amoxicillin and ciprofloxacin as antibiotics to demonstrate AST performance.
2. A stock solution of oxygen-sensing nanoprobes (Oxnano, PyroScience GmbH): dispense the probe particles (10 mg) in 10 mL of DI water. Use ultrasonication to break down small clusters of particles.
3. Amoxicillin stock solution: prepare 10 mL of the 2 mg/mL amoxicillin solution dissolved in phosphate buffer, pH 6. Make sure that all the powder is properly dissolved. Pipette (~250 μL) of the stock solution in the multiple 1.5 mL centrifugation tubes and store them at −20 °C. Use each aliquot of the stock solution only once, without refreezing.
4. Ciprofloxacin stock solution: prepare 10 mL of a 2.5 mg/mL ciprofloxacin hydrochloride monohydrate stock solution in DI water. Pipette (~250 μL) of the stock solution in the multiple 1.5 mL centrifugation tubes and store them at −20 °C. Use each aliquot of the stock solution only once, without refreezing.
5. Gel aliquots: Weight 0.06 g of the ultralow gelling temperature agarose per 1 mL of MHB II, in a 1.5 mL centrifugation tube. Heat the gel to 85 °C in a thermomixer until the gel is fully dissolved. Use centrifugation (60 s, 3700 rpm) to remove the

bubbles from the solution. Split the solution into 100 μL aliquots in separate centrifugation tubes using a positive displacement pipette while maintaining the temperature of the solution at 85 °C (*see* **Note 3**).

2.3 Preparation of the Microchamber Array Plate

The arrangement and design of the chamber array are drawn with CAD software (AutoCAD, Autodesk) and printed on a foil mask with a high-resolution printer. The preparation of a microchamber array plate requires clean-room equipment such as a plasma treatment system, a spin-coater, a hot plate, a mask aligner, and a nanoimprinting tool (CNI V2.0, NILT).

1. Silicon wafers, 4-inches.
2. Glass wafers, 4-inches.
3. Photoresist (SU-83050) with the corresponding developer solution (Mr-Dev 600).
4. COC foil (grade 8007 thickness of 240 μm), cut into 4-inch-size circles.
5. Trichloro(1H,1H,2H,2H-perfluorooctyl) silane (PFOTS) for the surface modification.
6. Priming solution (Ormoprime).
7. UV curable resist (Ormostamp).
8. Polydimethylsiloxane (PDMS), monomer and curing agent.
9. Polytetrafluoroethylene (PTFE) foil (2×)—protective layer for the nanoimprinting tool.

2.4 Fabrication of the Clamping Device

To perform measurements of the changes in the oxygen consumption by bacterial cells, it is important to maintain chambers with cells enclosed during the entire cultivation time. For this reason, a clamping device applying slight pressure on the glass slide and chamber array plate is necessary. Here, we present an example of the clamping device closed by a set of magnets. The design of the clamping system was conceptualized in CAD software, transferred to "stl" file format, and manufactured with a 3D printer (Ultimaker^3) from acrylonitrile butadiene styrene (ABS) material. The device was prepared in two parts (Fig. 2). The bottom part was designed to fit a 96-wellplate microscopy adapter. The bottom part has an opening that precisely fits a glass slide (24 × 40 mm^2, #5). The top part fits the bottom part so that the glass slide and the COC microchamber array plate are in contact during the measurements. To ensure tight contact between these parts, we incorporated openings for eight magnets of different sizes. The top part also contained a cut-out block of cured polydimethylsiloxane (PDMS, Sylgard 184) supported by a laser-cut poly(methyl methacrylate) (PMMA) block (height: 2 mm), shaped in the size that fits the window in the 3D printed top part of the clamping

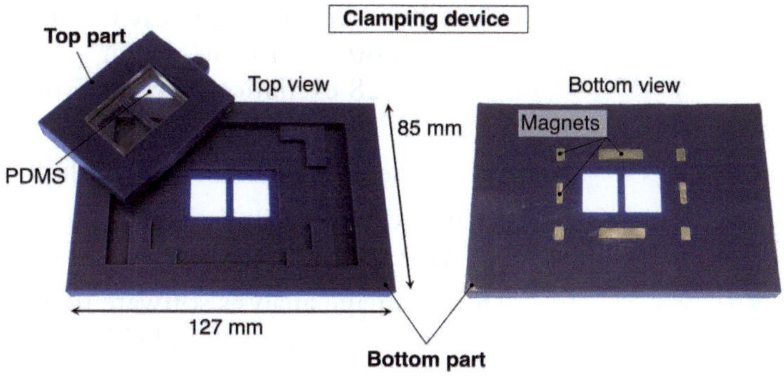

Fig. 2 Clamping device that ensures a tight contact between the chamber array and glass slide during the measurements. Both 3D-printed parts contain slots with glued magnets. The glass slide is overlayed with the chamber array system and placed in the center of the bottom part of the clamping device. The tight connection between the glass slide and the chamber system is enabled by a block of a flexible polymer, PDMS, which is inside the top part of the clamping device and supported on the top by a slice of PMMA. Both parts have an opening in the center for microscopy observation in bright field and fluorescence mode

device (22.4×38.3 mm^2) with a laser-cutter. The PDMS block evenly distributes the applied pressure by the magnets over the whole chamber system.

1. Block magnets (in this work: four $20 \times 5 \times 2$ mm^3; four $10 \times 3 \times 2$ mm^3; eight $6 \times 4 \times 2$ mm^3, obtained from supermagnete.ch) are glued inside the 3D printed clamping device.
2. PDMS block: Mix the PDMS monomer and curing agent in a ratio of 10:1. Pour the mixture into a flat-bottom container, for example, a square petri dish, place it on the well-leveled surface inside the oven, and cure at 80 °C overnight. Once PDMS is polymerized, remove it from the container and cut in the pieces corresponding to the opening in the top part of the clamping device. The leftover PDMS will be used in the following part of the protocol for filling the microchamber system (*see* **Note 4**).

2.5 Imaging Setup and Data Analysis

The imaging of the microchamber array plate is performed on a motorized, inverted wide-field microscope (Nikon Ti-E, equipped with NIKON NIS-Elements Advanced Research software).

1. Environmental box to maintain the temperature of the whole system at 37 °C.
2. Light source for the excitation of the oxygen-sensing nanoprobes and produced sfGFP: Lumencor Spectra X LED (30% blue LED intensity, exposure time: 200 ms) and (30% cyan LED, exposure time: 100 ms), respectively.

3. Filter set: 438/24 excitation filter, 660 dichroic mirror, 785/60 emission filter (oxygen-sensing nanoprobes); 475/28 excitation filter, 495 dichroic, 525/50 emission filter (sfGFP).
4. Camera: Hamamatsu Orca Flash 4.
5. Objective: 10× objective: Nikon, Plan Fluor NA:0.3, WD:16 mm.
6. Fiji image analysis software to analyze luminescence intensities from the images of individual chambers [27].

3 Methods

3.1 Fabrication of the Chamber Array Plate

The chamber array plate was fabricated from the air-tight material, COC by thermal imprinting (hot embossing). The process involves the preparation of a silicon master mold with the array design of a selected height and transferring this design to a COC foil by thermal imprinting. **Steps 1–7, 10–12**, and **14** are performed in a clean room environment.

1. To prepare the initial master mold, plasma-treat a 4-inch silicon wafer (200 sccm argon, 40 sccm oxygen, 200 W, 3 min).
2. Spin-coat photoresist (SU-83050) on the silicon wafer at 2750 rpm for 30 s, resulting in a structure height of 75 μm.
3. Soft bake the wafer at 95 °C for 30 min.
4. Expose the wafer with UV light (200 mJ/cm^2, i-line) through a photomask on a mask aligner.
5. Perform a post-exposure bake of the wafer at 95 °C for 4 min on a hot plate.
6. Develop the exposed photoresist by immersing the wafer in a SU-8 developer solution (Mr-Dev 600) for approximately 10 min. Agitation helps to facilitate photoresist removal from the areas between the chambers.
7. Hard bake the wafer for 1 h at 160 °C.
8. To avoid adhesion of the master wafer in the following fabrication process, perform a silanization of the wafer. Place the wafer in the desiccator together with 100 μL of trichloro (1H,1H,2H,2H-perfluorooctyl) silane (PFOTS) overnight under a reduced pressure of 100 mbar.
9. Mix and degas 40 g of PDMS base polymer and curing agent in a 10:1 ratio and pour it on the SU-8 mold. After curing the PDMS for 1 h at 80 °C, detach and cut the PDMS to a 4-inch format. The PDMS is also silanized as described in **step 8**.

10. Transfer the design from the PDMS mold to the stamp used for the thermal imprinting. A 4-inch glass wafer is plasma-treated for 3 min using conditions as in **step 1**. Next, the glass wafer is primed by spin coating 3 mL of a priming solution (Ormoprime) at 4000 rpm for 1 min. Subsequently, the wafer is baked at 150 °C for 5 min.
11. Dispense 4 mL of a UV curable resist (Ormostamp) on the glass wafer. Next, the resist is sandwiched with the silanized PDMS mold. Subsequently, the stack is cured in a UV-exposure box for 3 min, transferring the structures onto the glass wafer, referred to as the stamp.
12. The resulting stamp is hard-baked at 130 °C for 30 min and also silanized to avoid adhesion of the stamp to the COC foil; repeat **step 8**.
13. Before starting the thermal imprinting process, cut out a 4-inch wafer-sized piece from the COC foil (grade 8007 thickness of 240 μm).
14. Place a stack of PTFE foil, a bare silicon wafer, the COC foil, the stamp, and another PTFE foil into the nanoimprinting tool (CNI V2.0, NILT). By applying a pressure of 6 bar for 4 min at 130 °C, the design is transferred from the stamp to the foil. A clean silicon wafer (4 inches) is used during the imprinting to obtain a smooth, thereby transparent back side of the imprinted COC foil.
15. Cut the imprinted COC foil to separate the individual chamber array systems (*see* **Note 5**).

3.2 Bacterial Cultivation

1. Sterilize the surface of the bottle containing MHB II medium, 2 mL serological pipette and 14 mL cultivation tube. Place the materials inside the sterile cabinet.
2. Transfer 2 mL of the medium to the cultivation tube and add antibiotics for the plasmid maintenance to a final concentration of 50 μg/mL kanamycin sulfate and 100 μg/mL ampicillin sodium salt).
3. Thaw an aliquot of the cryo-stock (*see* **Note 6**) and pipette 25 μL of the cryo-stock to the prepared cultivation tube.
4. Place the cultivation tube with the inoculated medium in the incubation shaker at 37 °C and 220 rpm.
5. Wait until the turbidity of the inoculated medium reaches OD_{600} value between 0.4 and 0.6 and dilute the culture in the fresh medium containing only kanamycin sulfate to obtain 0.5 mL with OD_{600} value of 0.02 and use it for the experiment.

3.3 Performing AST

1. Select four centrifugation tubes with the 100 μL aliquots and place them into a thermomixer at a temperature of 85 °C. Once the temperature is reached and the gel is fully melted, add to each tube 20 μL of the 1 mg/mL solution of OXNANO oxygen-sensing nanoprobes (*see* **Note 7**).
2. Heat the gel with particles to 95 °C and hold it at a temperature for at least 5 min to prevent potential contamination from the previous manipulation steps.
3. Let the tubes cool down to 47 °C while preparing the antibiotic solutions. The antibiotic solutions are prepared in MHB II medium (2× the target concentration) based on the EUCAST recommendation for the stock solutions solvents. Heat the stock solutions of antibiotics (and medium, for the control) to 47 °C, and add 120 μL of the solution to the gel with nanoprobes. Mix the content of the tube using a vortex and return to the thermomixer.
4. Prepare the plasma-activated COC array (oxygen plasma, 0.65 mbar, 18 W, 45 s) and place it on a clean surface. Prepare four thin slices of cured PDMS or any different flexible material.
5. Clean a glass slide and place it in the 3D-printed clamping device.
6. Keep the mixture of gel, particles, and antibiotics at 47 °C and pipet subsequently 20 μL over each array. Spread the gel using a small piece of PDMS. The mixture cools down in the chambers and solidifies almost instantly. Due to the evaporation of the liquids, this step needs to be performed as quickly as possible. The extensive evaporation of the liquid from the chambers leads to the formation of bubbles and inhomogeneities in the cell inoculation.
7. Pipette four aliquots (5 μL) of the bacterial culture on a cleaned glass slide (*see* **Note 8**). Cover the glass slide with the array containing gel so that each droplet of bacterial culture is below the selected array of chambers. **Steps 6** and 7 are depicted in Fig. 3.
8. Close the clamping device to maintain the glass slide and bacterial culture in contact over the measurement.
9. Place the clamping device on the stage of the motorized microscope and proceed with the data acquisition.

3.4 Data Acquisition and Analysis

The images are taken in the bright field and epi-fluorescence mode during excitation by the LED. Acquired images are used to determine changes in the oxygen level in the chambers during the cultivation period. The bright field images are used to study changes in the morphology due to antibiotic exposure; they can serve as a control; however, they can be omitted from the AST measurement.

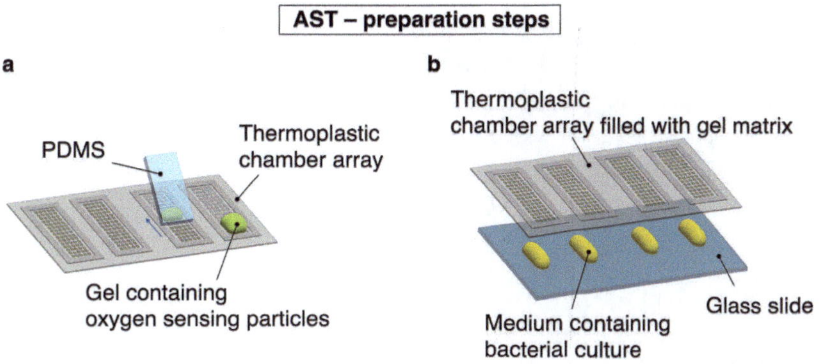

Fig. 3 Schematic representation of the AST protocol. (**a**) Pipette approximately 20 μL of the gel solution and spread the gel into the plasma-activated COC chamber plate using a thin PDMS piece. (**b**) Pipette 5 μL of the diluted bacterial culture on a clean glass slide and overlay it with the prepared chamber plate

1. Select chambers for recording and set up the data acquisition, for example, every 15 min during the 5-h time period (*see* **Note 9**).
2. Open the Fiji image analysis software.
3. Open the obtained file with the time-lapse images collected during the cultivation. Use split channels function during the import so that the bright filed and luminescence images are open as separate windows, containing all the different time points in a single stack.
4. Adjust the brightness and contrast of the image to clearly distinguish individual chambers on the images. Open the ROI manager inside the software. Using the ROI manager, specify the size of four ROIs, 315 × 315 μm, placed around the four chambers and add these ROIs to the ROI manager. An example of the four chambers in the epifluorescence channel with the marked ROI is in Fig. 4a.
5. Set the measurement inside the Fiji software to obtain area, mean grey value, and integrated density for each chamber. Using ROI manager, perform the "Multi Measure" command and collect the values in the form of a table.
6. Generate a table containing all the mean grey values separately for each condition.
7. From all the series of time-resolved data points, subtract the mean grey value at the time equal to zero minutes (*see* **Note 10**).
8. Plot the collected data separately for each condition with a data analysis software such as Origin. An example of the obtained curve for the *E. coli* ATCC 35218 *[pSEVA271_sfgfp]* is in Fig. 4b. Here, we also used images in bright field and produced sfGFP to correlate measured changes in the luminescence of

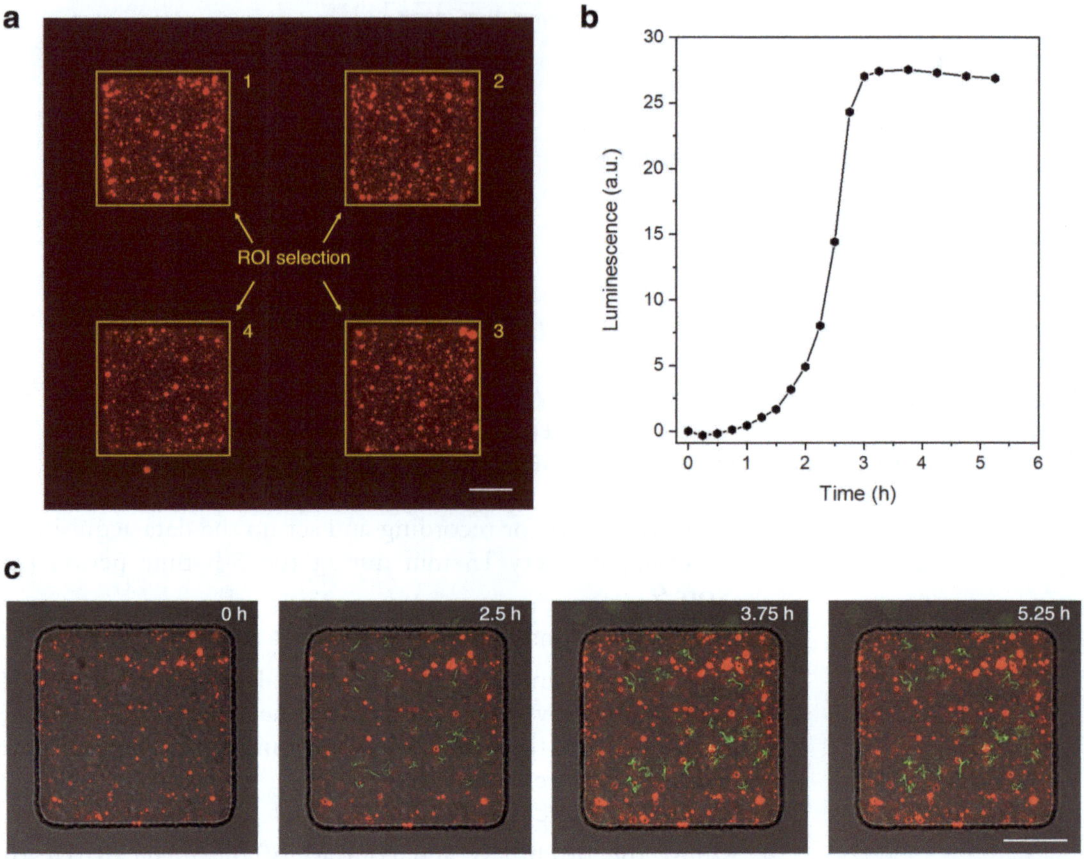

Fig. 4 (**a**) Data analysis with Fiji software. Set of four chambers open as epifluorescence channels. The area of all chambers is surrounded by an ROI set in the ROI manager as 315 × 315 μm. The mean grey value is collected from all four areas. (**b**) An example of data obtained from the chamber area (marked as 3). The mean grey value is plotted as a function of cultivation time. The data are plotted after the subtraction of the mean grey value at the beginning of the cultivation as an initial oxygen level in the chamber. (**c**) Time-lapse images from the cultivation of the *E. coli* cells without antibiotic exposure, as a composite of the bright field and two epifluorescence channels (red: oxygen-sensing nanoprobes, green: bacterial cells). The images show the increase of the bacterial biomass and the following increase in the luminesce of the oxygen-sensing nanoprobes. Scale bars: 100 μm

oxygen-sensing nanoprobes with the observations of bacterial biomass increase and morphological changes. Figure 4c shows that for bacteria not exposed to antibiotics, biomass, as well as luminescence of the nanoprobes, increases during the cultivation time.

Similar results were obtained when we exposed *E. coli* ATCC 35218 *[pSEVA271_sfgfp]* to amoxicillin at a concentration 4 times greater than MIC_{ref} (4 μg/mL). In contrast, when we exposed cells to ciprofloxacin, at a concentration again 4 times greater than MIC_{ref} (0.008 μg/mL) we did not observe an increase in biomass (green) or change in the luminescence of

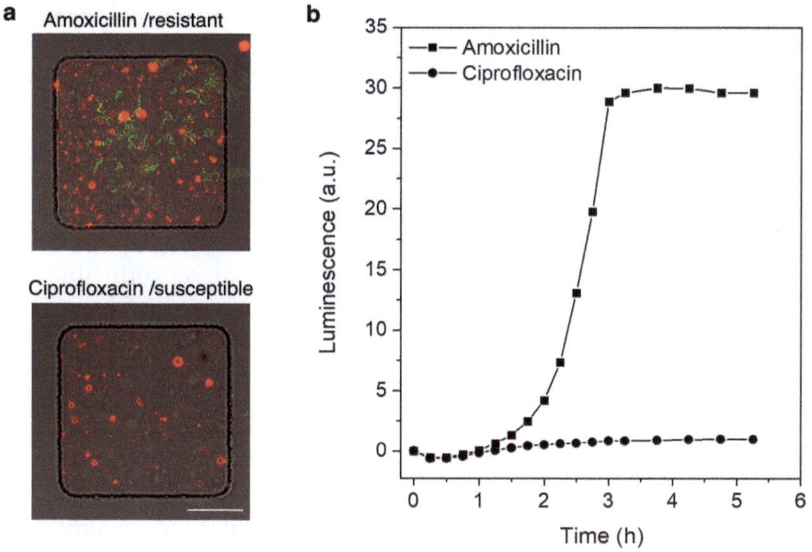

Fig. 5 Results of the AST testing: (**a**) The images showing the selected chambers with bacteria treated with amoxicillin and ciprofloxacin at the final point of the incubation (5.25 h) and the corresponding time-resolved curves (**b**) confirming the microscopy observation that the bacterial strain is resistant to amoxicillin and susceptible to ciprofloxacin. Scale bar: 100 μm

oxygen-sensing nanoprobes (red). MIC_{ref} for both antibiotics was taken from the reference [26] for *E. coli* ATCC 25922.The figures of selected chambers for each condition, together with the corresponding graph with time-resolved changes in oxygen consumption for these chambers, are in Fig. 5. Thereby, the system is suitable for determining the resistance/susceptibility of the studied bacterial cells.

4 Notes

1. The *E. coli* strain ATCC 35218 *[pSEVA271_sfgfp]* was prepared by transformation using electroporation [28] with the plasmid pSEVA271_sfgfp (lab collection) carrying the *sfgfp* gene. The *sfgfp* gene was constructed based on plasmid pSEVA271 [29] harboring kanamycin resistance by adding a gene for sfGFP [30] under the control of a constitutive promoter.
2. After autoclaving, the prepared medium can be stored as aliquots (10–50 mL at room temperature).
3. The aliquots can be used immediately or let to solidify at room temperature and stored in the fridge (4 °C) until use. Avoid repeated cycles of re-melting. Before use, place the tubes again in a thermomixer and heat to 85 °C.

4. The thickness of the cured PDMS should be slightly (~1 mm) higher than the opening in the top part of the clamping device with the already inserted PMMA block. To ensure this, we first prepared the PDMS block inside the container, and once we knew the thickness of the cured polymer, we 3D printed the top part of the device with the adjusted height. When we needed further PDMS blocks, we used the same volume of the PDMS monomer and the curing agent mixture, and we used the same container to ensure reproducibility of the PDMS height.
5. Typically, one wafer contains five copies of the array design.
6. The advantage of preparing bacterial cultures from the aliquots of cryo-stocks is the inoculation reproducibility, which is important in the characterization and validation part of the new technique and protocol. In the next stage, when performing AST with different than quality control strains, we advise either following clinical guidelines and selecting colonies directly from the agar plate or adjusting the bacterial cultivation to particular experimental requirements.
7. Place the tube containing the probe solution in an ultrasonic bath for 1 h before using it to disrupt small clusters of nanoparticles and achieve more homogenous signal during measurements.
8. The pipette tip can be used to slightly spread the droplet to cover a larger area of the glass slide.
9. Adjust the imaging period and frequency to doubling time of the studied bacterial strain. Here, we used 15-min periods for the first 195 min of cultivation time, followed by 30-min periods for the rest of the AST time. Chambers with damage to the gel or bubbles are typically excluded from the measurement or the subsequent analysis. Using the setup described here, the field of view contains four chambers; thereby, from each recorded position, there are four chambers that can be analyzed. NIKON NIS-Elements Advanced Research software allows images to be exported as individual files. Use Split Multipoints command to obtain all the time points separate for each position. This helps to avoid processing of large files during the data analysis.
10. Considering that there are some oxygen-sensing particles and cells in the areas outside the chambers, we do not use these areas for the background subtraction. Instead, we use the initial value of the luminescence collected for each individual chamber as the background.

References

1. Karlowsky JA, Lob SH, Raddatz J et al (2021) In vitro activity of imipenem/relebactam and ceftolozane/tazobactam against clinical isolates of gram-negative bacilli with difficult-to-treat resistance and multidrug-resistant phenotypes-study for monitoring antimicrobial resistance trends, United State. Clin Infect Dis 72: 2112–2120
2. Unemo M, del Rio C, Shafer WM (2016) Antimicrobial resistance expressed by Neisseria gonorrhoeae : a major global public health problem in the 21st century. Microbiol Spectr 4
3. Jorgensen JH, Ferraro MJ (2009) Antimicrobial susceptibility testing: a review of general principles and contemporary practices. Clin Infect Dis 49:1749–1755
4. Fluit AC, Visser MR, Schmitz FJ (2001) Molecular detection of antimicrobial resistance. Clin Microbiol Rev 14:836–871
5. Trinh TND, Lee NY (2021) Nucleic acid amplification-based microfluidic approaches for antimicrobial susceptibility testing. Analyst 146:3101–3113
6. Qin N, Zhao P, Ho EA et al (2021) Microfluidic technology for antibacterial resistance study and antibiotic susceptibility testing: review and perspective. ACS Sens 6:3–21
7. Baltekin Ö, Boucharin A, Tano E et al (2017) Antibiotic susceptibility testing in less than 30 min using direct single-cell imaging. Proc Natl Acad Sci USA 114:9170–9175
8. Zhang P, Kaushik AM, Hsieh K, et al (2021) A cascaded droplet microfluidic platform enables high-throughput single cell antibiotic susceptibility testing at scale 2101254:1–12
9. Kaminski TS, Scheler O, Garstecki P (2016) Droplet microfluidics for microbiology: techniques, applications and challenges. Lab Chip 16:2168–2187
10. Avesar J, Rosenfeld D, Truman-Rosentsvit M et al (2017) Rapid phenotypic antimicrobial susceptibility testing using nanoliter arrays. Proc Natl Acad Sci 114:E5787–E5795
11. Zeng W, Chen P, Li S et al (2022) Hand-powered vacuum-driven microfluidic gradient generator for high-throughput antimicrobial susceptibility testing. Biosens Bioelectron 205:114100
12. Yi Q, Cai D, Xiao M et al (2019) Direct antimicrobial susceptibility testing of bloodstream infection on SlipChip. Biosens Bioelectron 135:200–207
13. Liu X, Li X, Wu N et al (2022) Formation and parallel manipulation of gradient droplets on a self-partitioning SlipChip for phenotypic antimicrobial susceptibility testing. ACS Sens 7: 1977–1984
14. Lee W-B, Chien C-C, You H-L et al (2019) An integrated microfluidic system for antimicrobial susceptibility testing with antibiotic combination. Lab Chip 19:2699–2708
15. Li H, Torab P, Mach KE et al (2019) Adaptable microfluidic system for single-cell pathogen classification and antimicrobial susceptibility testing. Proc Natl Acad Sci 116:10270–10279
16. Wang H, Conover GM, Han S-I et al (2021) Development of single-cell-level microfluidic technology for long-term growth visualization of living cultures of mycobacterium smegmatis. Microsyst Nanoeng 7:37
17. Sun H, Chan C-W, Wang Y et al (2019) Reliable and reusable whole polypropylene plastic microfluidic devices for a rapid, low-cost antimicrobial susceptibility test. Lab Chip 19: 2915–2924
18. Azizi M, Davaji B, Nguyen AV et al (2021) Gradient-based microfluidic platform for one single rapid antimicrobial susceptibility testing. ACS Sens 6:1560–1571
19. Besant JD, Sargent EH, Kelley SO (2015) Rapid electrochemical phenotypic profiling of antibiotic-resistant bacteria. Lab Chip 15: 2799–2807
20. Spencer DC, Paton TF, Mulroney KT et al (2020) A fast impedance-based antimicrobial susceptibility test. Nat Commun 11:5328
21. Cira NJ, Ho JY, Dueck ME et al (2012) A self-loading microfluidic device for determining the minimum inhibitory concentration of antibiotics. Lab Chip 12:1052–1059
22. Ndieyira JW, Watari M, Barrera AD et al (2008) Nanomechanical detection of antibiotic-mucopeptide binding in a model for superbug drug resistance. Nat Nanotechnol 3:691–696
23. Etayash H, Khan MF, Kaur K et al (2016) Microfluidic cantilever detects bacteria and measures their susceptibility to antibiotics in small confined volumes. Nat Commun 7:1–9
24. Jusková P, Schmitt S, Kling A et al (2021) Real-time respiration changes as a viability indicator for rapid antibiotic susceptibility testing in a microfluidic chamber array. ACS Sens 6: 2202–2210

25. Clinical and Laboratory Standards Institute (CLSI) (2018) M100Ed30 | Performance standards for antimicrobial susceptibility testing, 28th edn
26. European Committee for Antimicrobial Susceptibility Testing (EUCAST) of the European Society of Clinical Microbiology and Infectious Diseases (ESCMID) (2000) Determination of minimum inhibitory concentrations (MICs) of antibacterial agents by agar dilution. Clin Microbiol Infect 6:509–515
27. Schindelin J, Arganda-Carreras I, Frise E et al (2012) Fiji: an open-source platform for biological-image analysis. Nat Methods 9: 676–682
28. Green MR, Sambrook J (2012) Molecular cloning: a laboratory manual. Cold Spring Harbor Laboratory Press
29. Silva-Rocha R, Martínez-García E, Calles B et al (2013) The standard European vector architecture (SEVA): a coherent platform for the analysis and deployment of complex prokaryotic phenotypes. Nucleic Acids Res 41: D666–D675
30. Pédelacq J-D, Cabantous S, Tran T et al (2006) Engineering and characterization of a superfolder green fluorescent protein. Nat Biotechnol 24:79–88

Chapter 12

An Immiscible-Phase Filtration Device for Isolation, Amplification, and Detection of Nucleic Acids for Clinical Diagnostics

Bongkot Ngamsom, Pablo Rodriguez-Mateos, and Nicole Pamme

Abstract

Clinical diagnostics of infectious diseases via nucleic acid amplification tests (NAATs) depend on a separate step of isolation of nucleic acids from cells/viruses embedded in complex biological matrices. The most recent example has been reverse transcription polymerase chain reaction (RT-PCR) for amplification and detection of SARS-CoV-2 RNA for COVID-19 diagnostics. Kits for RNA extraction and purification are commercially available; however, their integration with amplification systems is generally lacking, resulting in two separate steps, i.e., sample preparation and amplification. This makes NAATs more time-consuming, requiring skilled personnel, and can increase the likelihood of contamination. Here, we describe a setup and methodology to perform the quick extraction and detection of nucleic acids in an integrated manner. In particular, we focus on the use of an immiscible filtration device for capture, isolation, concentration, amplification, and colorimetric detection of SARS-CoV-2 RNA.

Key words IFAST, COVID-19, Immiscible filtration, Extraction, Isothermal amplification, Diagnostics

1 Introduction

Owing to their high sensitivity and specificity, nucleic acid amplification tests (NAATs) have become the preferred methods for many clinical diagnostics, e.g., severe acute respiratory syndrome coronavirus 2 (SARS-CoV-2) [1, 2], human immunodeficiency virus (HIV) [3, 4], tuberculosis [5, 6], *Chlamydia trachomatis* and *Neisseria gonorrhoeae* infections [7, 8]. These tests involve a sample preparation step to extract and purify nucleic acids (DNA or RNA) from complex biological matrices (e.g., blood, urine, feces, nasopharyngeal/oropharyngeal swabs, and genital track swabs), prior to performing amplification and detection of the amplicons to obtain results. Although commercially available kits can be used, multiple pipetting, centrifugation, and washing steps are usually

Valérie Taly et al. (eds.), *Microfluidics Diagnostics: Methods and Protocols*, Methods in Molecular Biology, vol. 2804,
https://doi.org/10.1007/978-1-0716-3850-7_12,

required [9]. This time-consuming, labor-intensive, and contamination-prone process prohibits the full potential use of NAATs for clinical diagnostics, especially in resource-constrained settings.

Utilizing microscale gates in combination with the surface tension properties of immiscible liquids (e.g., oil-aqueous), the immiscible filtration assisted by surface tension (IFAST) method offers an elegant way to perform sequential and multiple assays in one single device [10, 11]. Comprising a larger sample chamber (0.5–1 mL) and a series of smaller wash chambers (20–120 μL) interconnected via microscale trapezoidal gates (100–200 μm deep), fluidic barriers between immiscible liquids can be created, allowing side-by-side liquid compartmentation in credit-card size devices [10–16]. Filtration of analytes of interest from complex matrices within the sample chamber is achieved by the use of a permanent magnet to manipulate functionalized superparamagnetic particles across the immiscible barriers into a separate aqueous solution [12]. This rapid and robust microscale purification (< 1 min) eliminates extra essential steps for extraction of analytes from biological samples prior to detection [11].

IFAST has seen numerous applications in sample preparations for specific isolation of nucleic acids [10, 11, 13, 17–19], bacteria [14, 15], and rare cells [12] from various biological samples. To take advantage of the IFAST platform further, a detection chamber within the same device has been introduced after purification steps to allow for sample-in to answer-out analyses. Our group combined the specificity of the IFAST with the sensitive ATP assays for detections of *Escherichia coli* from wastewater [15], and *Streptococcus agalactiae* from urine samples [14], allowing rapid detection of bacteria within 20 min. This provides a promising platform for point-of-need pathogen analyses to replace the gold standard cell culture method which requires 24–48 h to obtain results. Additionally, all-in-one IFAST devices for extraction, amplification, and detection of nucleic acids were also reported, exploiting the commercially available isothermal amplification kits for nucleic acids by Shaw's group for DNA identification of rare rhinoceros species [19], and by our group for SARS-CoV-2 RNA detection [11]. These robust platforms enable sample preparation and nucleic amplification tests to be integrated in one device, facilitating point-of-need/point-of-care testing in resource-limited settings.

Here, we discuss the application of the microfluidic IFAST platform as an example of an all-in-one device for RNA isolation, amplification, and detection for COVID-19 diagnostics (Fig. 1). However, the device is also largely applicable to other nucleic amplification assays for various pathogen analyses.

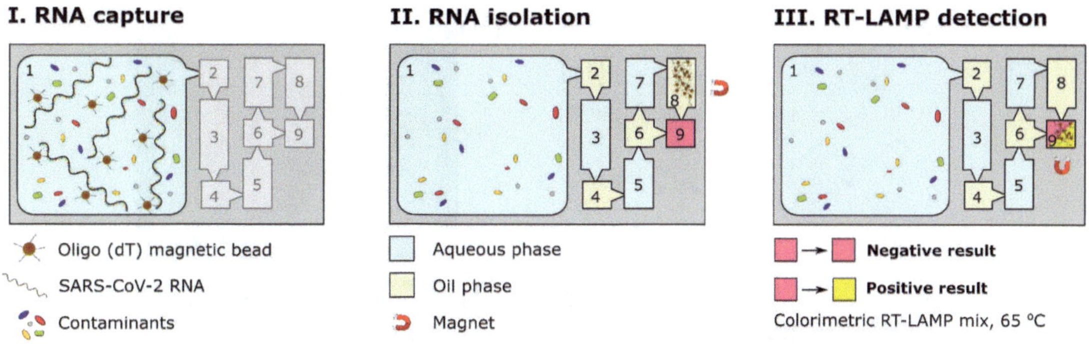

Fig. 1 Schematic demonstrating on-chip workflow for RNA isolation and detection utilizing immiscible liquids and functionalized magnetic particles, allowing sequential steps to take place in one device

2 Materials

Prepare all reagents related to nucleic acid amplification using nuclease-free water, and wash buffer and reagents for gel electrophoresis with purifying deionized water with a sensitivity of 18 MΩ-cm at 25 °C. Prepare and store all reagents at room temperature unless otherwise stated. Reagents purchased should be stored following the assay instruction manuals. Waste disposal regulations should be diligently followed when disposing waste materials.

2.1 Biomolecules, RT-LAMP Reagents, Magnetic Particles, and Immiscible Liquids

1. Genomic RNAs: SARS-CoV-2 RNA (2019-nCoV/USA-WA1/2020, ATCC VR-1986D), (LGC standards, UK) (*see* **Note 1**).
2. RT-LAMP assay kit: WarmStart® DNA and RNA polymerase (MS1800S, NewEngland Biolabs) (*see* **Note 2**).
3. LAMP primers (Integrated DNA Technology) (*see* **Note 2**).
4. Superparamagnetic particles: Dynabeads Oligo dT(25) (Invitrogen, UK) (*see* **Note 3**).
5. Wash buffer: Phosphate buffer saline solution containing 0.005% Tween20 (PBST). Dissolve a PBS tablet in 200 mL water. Add 2.5%w/v Tween 20 to afford a final concentration of 0.005%w/v. Store at 4 °C (*see* **Note 4**).
6. Mineral oil (*see* **Note 5**).

2.2 Microfluidic Chip Design and Fabrication

1. In a suitable design software (e.g., AutoCAD, SolidWorks), reproduce the chip design shown in Fig. 2a. The chip design features a large sample chamber 1 (26 mm wide, 26 mm long); wash chambers 2, 4, 6 (3 mm wide, 3 mm long); wash chamber 3 (3 mm wide, 14 mm long); wash chambers 5, 7, 8 (3 mm wide, 8.5 mm long); and a detection chamber 9 (3 mm wide,

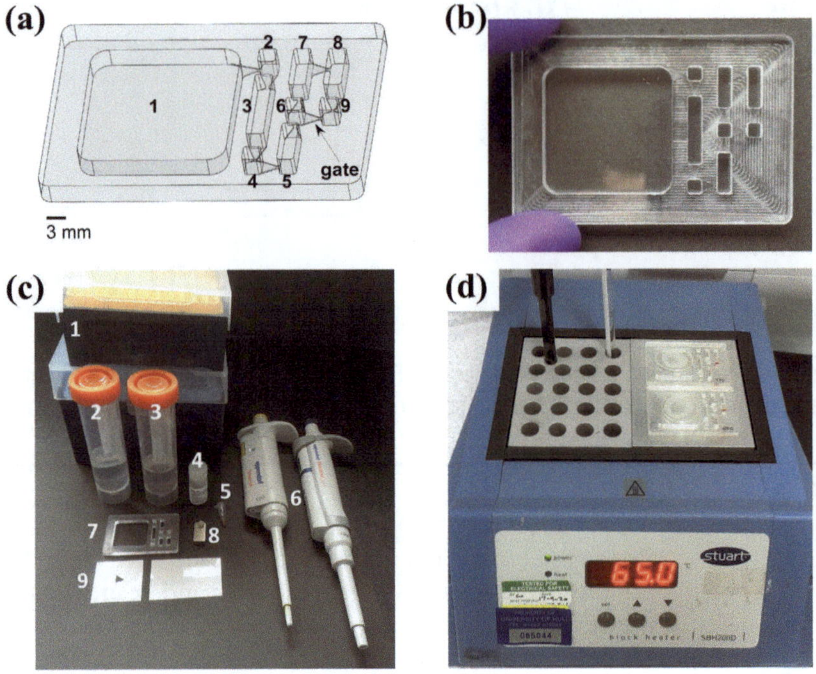

Fig. 2 (**a**) Design of the microfluidic device, featuring a large sample chamber (1); wash chambers (2–8), and a detection chamber (9), interconnected via gates. (**b**) Photograph of the device fabricated in PMMA. (**c**) All elements required for on-chip assays: 1 = micropipette tips, 2 = 0.005% Tween 20, 3 = mineral oil, 4 = oligo d(T) magnetic beads, 5 = RT-LAMP substrate, 6 = micropipettes, 7 = PMMA device, 8 = magnet assembly, 9 = adhesive tape. (**d**) A hotplate set at 65 °C was used as a heating source for on-chip assay. (Reprinted from Rodriguez-Mateos et al. [11], with permission from Elsevier)

3 mm long). All chambers were interconnected via gates (3 mm to 0.5 mm wide, 3 mm long).

2. Fabricate the chip design by CNC machine milling (Daltron M7, Milton Keynes) using a polymethyl methacrylate (PMMA) sheet (Kingston Plastics), to a depth of 3.8 mm for all chambers, and 0.2 mm for the gates (Fig. 2b) (*see* **Notes 6** and 7).

2.3 On-Chip Assay Setup

Figures 2c, d illustrate the elements required for on-chip assays.

1. Adhesive film for device sealing: Adhesive PCR plate seals (Thermofisher Scientific).
2. Neodemium-iron-boron (NdFeB) magnet (Magnet Sales) (*see* **Note 8**).
3. Guanidine hydrochloride (GuHCl) solution (Sigma-Aldrich): Dissolve GuHCl in PBST to afford a final concentration of 5 M. Keep at room temperature (*see* **Note 9**).

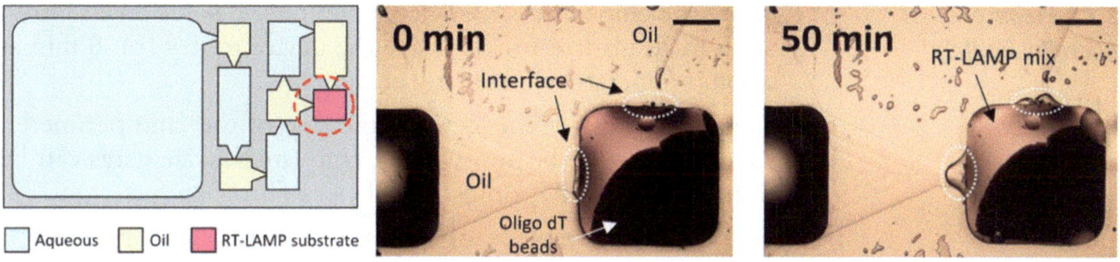

Fig. 3 Microscope images of interfaces between mineral oil and RT-LAMP reagents at 65 °C at 0- and 50-min incubation time. Scale bar = 1 mm. (Reprinted from Rodriguez-Mateos et al. [11], with permission from Elsevier)

4. Inverted microscope (e.g., Nikon Eclipse Ti) with a suitable camera (e.g., Nikon DS-Fi3) and image acquisition software (e.g., NIS-Elements Freeware (F)) (*see* **Note 10**, Fig. 3).
5. RNase surface decontamination solution: RNase Away Surface Decontaminant (Thermo Scientific) (*see* **Note 11**).
6. Blu-Tack (Bostik).
7. Heating source, e.g., hot plate (Stuart) or incubator.

2.4 Agarose Gel Electrophoresis

1. Tris Acetate-EDTA (TAE) buffer: Dilute 20 mL of 50× TAE buffer stock (Sigma-Aldrich) into 980 mL water. Store at room temperature.
2. Agarose gel: add agarose (Melford) into 1× TAE buffer to afford a final concentration of 1%w/v.
3. DNA gel stain: Add appropriate volume of 10,000× SYBR-safe DNA gel stain (Invitrogen) into melted agarose prior to casting.
4. DNA loading dye (DNA Gel Loading Dye, Thermofisher Scientific).
5. DNA ladder (GeneRuler DNA Ladder Mix, Thermofisher Scientific).
6. DNA electrophoresis setup (Mini-Sub Cell GT Horizontal Electrophoresis System and PowerPac Basic Power Supply, Bio-Rad).
7. Molecular imager (Chemidoc XRS+, BioRAD).
8. Cooking microwave.

3 Methods

It is recommended that the preparation of microfluidic chip (*see* Subheading 3.1), magnetic beads (*see* Subheading 3.2), and RT-LAMP reagents (*see* Subheading 3.3) is performed in a pre-amplification area, separated from the amplification and post-amplification area (*see* **Note 12**).

3.1 Preparation of Microfluidic Chip

1. Wipe the chip with RNase decontamination solution and rinse with nuclease-free water. Leave the device to dry (ca. 5 min at room temperature) prior to use.
2. Cut the adhesive film to the same size of the chip perimeter. Gently press the bottom side of chip (microscale gates can be seen from this side) onto the adhesive film.
3. Prepare another piece of adhesive film (ca. 28 mm wide × 28 mm long) with pre-cut hole (ca. 3 mm in diameter) in the middle to cover the sample chamber.

3.2 Preparation of Magnetic Beads

1. Resuspend the magnetic particles in the stock bottle either by shaking or using a vortex mixer.
2. Pipette required volume (20 μL per assay, *See* **Note 13**) from bead stock into a 1.5 mL Eppendorf tube.
3. Attach a permanent magnet to the side of the Eppendorf tube using Blu-Tack. Allow 2 min for the magnetic particles to be pulled to the side of the tube.
4. While the magnet is still in place, remove supernatant using a pipette.
5. Add 1 mL PBST to the beads. Agitate/Vortex for 5 s.
6. Attach a NdFeB magnet to the side of the Eppendorf tube using Blu-Tack.
7. While the magnet is still in place, remove supernatant using a pipette.
8. Suspend the washed beads in PBST to the volume originally taken out from the stock bottle.

3.3 Preparation of RT-LAMP Reagents

1. Reconstitute the primers in nuclease-free water according to the instruction manual (*see* **Note 2**).
2. Follow the instruction manual to prepare RT-LAMP master mix; make up total volume of 20(x + 1) μL, where x = number of assays to be performed, and one assay requires 20 μL reagent. The master mix contains 16 mM forward inner primer (FIP) and backward inner primer (BIP), 2 mM forward outer primer (F3) and backward outer primer (B3), 4 mM forward loop primer (LF) and backward loop primer (LB), 1× LAMP reaction mix and nuclease-free water.

3.4 On-Chip RNA Capture, Purification, Concentration, Amplification, and Detection (RT-LAMP)

1. Add 2 μL mineral oil to chamber 2.
2. Add 974 μL of 5 M GuHCl in PBST into the sample chamber followed by 25 μL washed Oligo (dT) magnetic beads and gently agitate to disperse magnetic beads inside the liquid.

3. Add 1 μL of RNA into the mix to afford the desired concentration. Always perform on-chip control experiments where RNA is replaced by nuclease-free water.
4. Place a small piece of adhesive film with a small pre-cut hole above the sample chamber to prevent spillage. Agitate the device for 10 min (*see* **Note 14**).
5. Alternately fill the wash chambers (2–8) with immiscible liquids: chambers 2, 4, 6 = mineral oil (20 μL) mineral oil; chamber 8 = mineral oil (50 μL); chamber 3 = PBST (100 μL); chambers 5, 7 = PBST (50 μL). Fill the detection chamber (9) with 20 μL of RT-LAMP reagent (see **Note 15**).
6. Use a permanent magnet (*see* **Note 8**) to gather RNA-extracted magnetic beads in the sample chamber toward the gate to chamber 2.
7. Use the permanent magnet to move the beads from the sample chamber (1), through wash chambers (2–8), into chamber 9 (*see* **Notes 16–18**).
8. Add 10 μL mineral oil onto RT-LAMP reagent in chamber 9. Place the chip device on ice and transfer the chip to the post-amplification area (*see* **Note 19**).
9. Heat the device at 65 °C to initiate RT-LAMP reaction for 30–45 min using a preheated hotplate or incubator (*see* **Note 20**).
10. Remove the device from heating and allow to cool for result interpretation (*see* **Note 21**).
11. Observe color in the detection chamber: pink = negative amplification; yellow = positive amplification (Fig. 4a). The color of the RT-LAMP reagent in the control chip should remain pink.
12. Carefully pipette the amplification product from the RT-LAMP chamber (9) into a PCR tube, and keep on ice until ready for agarose gel electrophoresis.

3.5 Agarose Gel Electrophoresis for Verification of On-Chip Detection Results

1. Add agarose to TAE buffer to afford a concentration of 1%w/v.
2. Dissolve agarose using a microwave (*see* **Note 22**).
3. Prepare the gel box (base plate + surrounding rubber molds) and comb.
4. Wait for the melted agarose to cool to ca. 60–65 °C, and add SYBR Safe to the required volume.
5. Pour agarose solution containing SYBR Safe into the gel box, use small pipette tips to remove bubbles, and leave the liquid to set at room temperature. During this time, prepare samples to be analyzed, by adding appropriate volume of DNA loading dye into each sample and keep on ice.

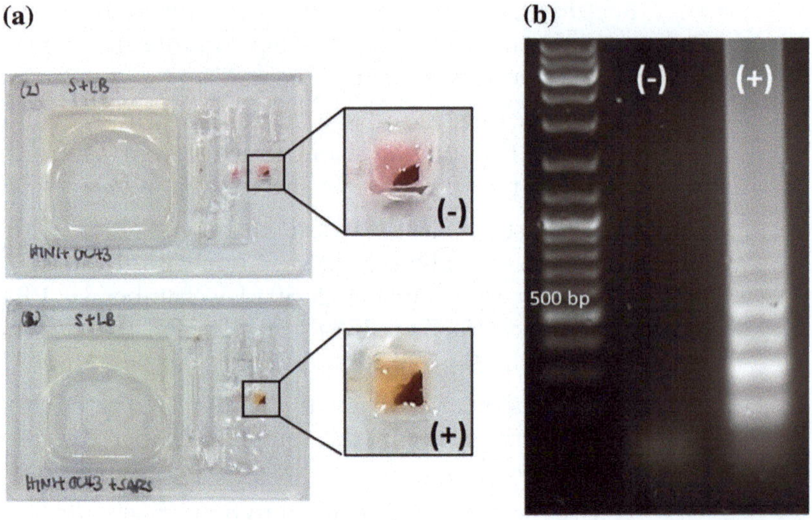

Fig. 4 (**a**) Negative (−) and positive (+) results from on-chip assays. (**b**) Confirmation of on-chip results by gel electrophoresis. (Reprinted from Rodriguez-Mateos et al. [11], with permission from Elsevier)

6. Once the gel is set, carefully remove the surrounding rubber molds to reveal the set gel on the base plate. Keep the comb inside the gel.
7. Carefully place the base plate carrying the gel into the gel tank pre-filled with appropriate volume of 1×TAE buffer. Gently remove the comb to reveal the wells into which amplification products will be loaded.
8. Use small pipette tips to load wells with DNA ladders (5 μL) and samples to be analyzed (20 μL per well).
9. Reattach the lid of the gel tank, and connect the power supply to the terminals on the lid. Set the voltage to 80 V and perform electrophoresis for 45 min.
10. Once finished, turn off the power supply. Remove the gel from the tank and wipe off excess liquid.
11. Image the gel (Fig. 4b) using a molecular imager.

4 Notes

1. Upon receipt, aliquot genomic RNAs to a volume required for each use and store at −80 °C. Avoid freeze-thaw cycles that can degrade RNA [20].
2. Follow instructions from manufacturers. Primer mix can be prepared in a large volume and aliquoted to a volume required for each use, and stored at −20 °C.

3. This work used Oligo dT(25) magnetic beads which selectively captured polyadenylated RNA species. However, they are fairly costly and require 2–8 °C storage. These magnetic beads can be replaced by the more economical silica paramagnetic particles. In this case, the selectivity of the on-chip assay will solely rely on the RT-LAMP primer specificity in the subsequent step.
4. Minute quantities of surfactant can be added to aqueous solutions to prevent sticking of magnetic beads to adhesive film used for sealing the chip, provided that the immiscible interfaces are not compromised. In our work, Tween 20 was added (final concentration of 0.005%w/v) in the sample chamber as well as wash buffer. Other surfactants, e.g., TritonX-100, can also be employed [21]; however, immiscible barriers between liquids should be checked under a microscope.
5. Mineral oil was employed as the immiscible phase. However, sunflower or olive oils among others can also be used, provided that they do not interfere with the downstream RT-LAMP assay (chamber 9). The quick way to check is to perform a tube-based assay overlaid with the oil of interest. These types of oils should provide cheaper and more accessible options for resource-constrained settings.
6. Polymer microfluidic devices can be fabricated via techniques such as hot embossing or injection molding to reduce the cost for mass production [22].
7. Polydimethylsiloxane devices with the same design were tested prior to PMMA devices. However, positive amplification to detect color change was longer than 60 min. PMMA devices allowed similar amplification to tube-based assays (30–40 min depending on initial RNA concentrations). Other thermoplastic materials such as polycarbonate can also be used.
8. We have found that a NdFeB magnet assembly worked well with our system. A bar magnet (20 mm × 10 mm × 5 mm) was used for gathering of beads in the sample chamber after incubation with RNA sample. A smaller disc magnet (4 mm diameter × 2 mm height) was used to manipulate the RNA-captured magnetic beads through the gates to the subsequent wash and RT-LAMP chambers. Other strong NdFeB magnets should also work.
9. The compatibility of GuHCl to on-chip assay [11] suggests the possible use for lysis of viral particles in real samples. While 5 M GuHCl (aq) was used in our investigation, the powder GuHCl can also be pre-stored inside the sample chamber [16].
10. The rapid and robust analyte isolation and purification utilizing this device relies on the filtration assisted by the surface tension properties of immiscible liquids. Therefore, it is crucial that the immiscible interfaces (oil aqueous) are kept intact to prevent

carry-over of the solution from the previous chamber into the next chamber. When first setting up the device and assay, monitor the placement and stability of the immiscible interfaces while considering the different buffer solutions, temperatures, and incubation times to be employed. This can be observed under an inverted microscope connected to a CCD camera. Figure 3 illustrates the immiscible interfaces between oil and RT-LAMP reagents during 65 °C from start up to 50 min.

11. Clean work surfaces prior to conducting experiments to eliminate RNase and DNA from laboratory surfaces using an RNase/DNA decontamination product.
12. It is advisable to perform experiments in two separate rooms for pre- and post-amplification steps to prevent DNA amplicon aerosol contamination. Change gloves regularly and avoid using the same pipettes for pre- and post-amplification steps.
13. The quantity of magnetic beads used for RNA isolation is dictated by (i) the gate size, which allows traverse of magnetic beads (too many beads can cause blockage in the gates) and (ii) capture efficiency (too few beads might lead to low capture efficiency and difficulty in visualization). The chip design herein can take up to 25 μL oligo(dT) magnetic beads (Thermofisher Scientific). Optimization is recommended for other types of beads.
14. When performing on-chip magnetic capture in the sample chamber, a small piece of adhesive film can be placed on top of the chamber to prevent spillages either during manual or automated mixing. Automated mixing can be achieved in a rotator (Stuart, UK; set at 40 rpm at 10–20° angle). The piece of adhesive tape also helps to avoid evaporation and movement of liquids during amplification. A small hole in the adhesive tape center helps to release building up of pressure during heating.
15. Fill the aqueous wash chambers (3, 5, 7) with solution before their oil counterparts (chambers 2, 4, 6, 8) to prevent oil penetration into adjacent chamber. Care should be taken when filling the RT-LAMP reagent chamber (9) as this contains some surfactant, that may result in overspill into the adjacent oil chambers. This can cause lower color intensity, and thus compromised result interpretation.
16. The integration of an on-chip amplification step after RNA isolation and purification eliminates the pipetting of beads-bound RNA into a separate amplification tube, reducing analyte loss and maintaining analyte integrity.

17. While particle loss can be expected during magnetic manipulation across chambers, this can be largely reduced by: (i) adding appropriate surfactant concentration, (ii) using a smaller magnet for passing through the gates, and (iii) maintaining a slow but steady movement, allowing beads to catch up with the magnet and move together as a group.
18. In rare occasions, small bubbles can appear when loading the aqueous and oil phases in the device chambers. Gently press the adhesive tape from the bottom side around the gate to move the bubble away into the chamber to rise to the surface.
19. The RT-LAMP reagent overlaid with mineral oil (10 μL) in chamber 9 is to prevent evaporation as well as contamination from DNA aerosol after amplification.
20. The block heater/incubator used for heating the on-chip RT-LAMP assay should be placed in the post-amplification area, and set at required temperature (65 °C) prior to performing the assay.
21. Ensure that color readouts are taken at room temperature. RT-LAMP color develops more strongly at lower temperature. Intensification of color can be achieved by placing the device on ice.
22. Care should be taken when preparing agarose gel using a microwave heating. Always use a container with volume ca. two times larger than actual volume to prevent spillage whilst heating.

References

1. Margolin E, Burgers W, Sturrock E et al (2020) Prospects for SARS-CoV-2 diagnostics, therapeutics and vaccines in Africa. Nat Rev Microbiol 18:690–704
2. Gandhi R, Lynch J, del Rio C (2020) Mild or moderate Covid-19. N Engl J Med 383:1757–1766
3. Sullivan T, Patel P, Hutchinson A et al (2011) Evaluation of pooling strategies for acute HIV-1 infection screening using nucleic acid amplification testing. J Clin Microbiol 49: 3667–3668
4. Emerson E, Plough K, Pangloli A et al (2013) Detection of acute HIV-1 infections utilizing NAAT technology in Dallas, Texas. J Clin Virol 58:E48–E53
5. MacLean E, Kohli M, Weber S et al (2020) Advances in molecular diagnosis of tuberculosis. J Clin Microbiol 58:10-1128
6. Acharya B, Acharya A, Gautam S et al (2020) Sapkota, advances in diagnosis of tuberculosis: an update into molecular diagnosis of mycobacterium tuberculosis. Mol Biol Rep 47:4065–4075
7. Hardick A, Hardick J, Wood B et al (2006) Comparison between the gen-probe transcription-mediated amplification trichomonas vaginalis research assay and real-time PCR for trichomonas vaginalis detection using a Roche LightCycler instrument with female self-obtained vaginal swab samples and male urine samples. J Clin Microbiol 44:4197–4199
8. Cosentino L, Campbell T, Jett A et al (2012) Use of nucleic acid amplification testing for diagnosis of anorectal sexually transmitted infections. J Clin Microbiol 50:2005–2008
9. Emaus M, Varona M, Eitzmann D et al (2020) Nucleic acid extraction: fundamentals of sample preparation methodologies, current advancements, and future endeavors. Trac-trends. Anal Chem 130:115985
10. Berry S, Alarid E, Beebe D (2011) One-step purification of nucleic acid for gene expression

analysis via immiscible filtration assisted by surface tension (IFAST). Lab Chip 11:1747–1753
11. Rodriguez-Mateos P, Ngamsom B, Walter C et al (2021) A lab-on-a-chip platform for integrated extraction and detection of SARS--CoV-2 RNA in resource-limited settings. Anal Chimi Acta 1177:338758
12. Berry S, Strotman L, Kueck J et al (2011) Purification of cell subpopulations via immiscible filtration assisted by surface tension (IFAST). Biomed Microdevices 13:1033–1042
13. Strotman L, Lin G, Berry S et al (2012) Facile and rapid DNA extraction and purification from food matrices using IFAST (immiscible filtration assisted by surface tension). Analyst 137:4023–4028
14. Ngamsom B, Wandera E, Iles A et al (2019) Rapid detection of group B streptococcus (GBS) from artificial urine samples based on IFAST and ATP bioluminescence assay: from development to practical challenges during protocol testing in Kenya. Analyst 144:6889–6897
15. Ngamsom B, Truyts A, Fourie L et al (2017) A microfluidic device for rapid screening of E-coli O157:H7 based on IFAST and ATP bioluminescence assay for water analysis. Chem Eur J 23:12754–12757
16. Mosley O, Melling LM, Tarn MD et al (2016) Sample introduction interface for on-chip nucleic acid-based analysis of helicobacter pylori from stool samples. Lab Chip 16:2108–2115
17. Strotman L, O'Connell R, Casavant B et al (2013) Selective nucleic acid removal via exclusion (SNARE): capturing mRNA and DNA from a single sample. Anal Chem 85:9764–9770
18. Berry S, LaVanway AJ, Pezzi HM et al (2014) HIV viral RNA extraction in wax immiscible filtration assisted by surface tension (IFAST) devices. J Mol Diagn 16:297–304
19. Wimbles R, Melling LM, Cain B et al (2021) On-site genetic analysis for species identification using lab-on-a-chip. Ecol Evol 11:1–9
20. Kellman B, Baghdassarian H, Pramparo T et al (2021) Multiple freeze-thaw cycles lead to a loss of consistency in poly(A)-enriched RNA sequencing. BMC Genomics 22:1–15
21. Berry SM, Regehr KJ, Casavant BP et al (2013) Automated operation of immiscible filtration assisted by surface tension (IFAST) arrays for streamlined analyte isolation. J Lab Autom 18: 206–211
22. Becker H, Gartner C (2008) Polymer microfabrication technologies for microfluidic systems. Anal Bioanal Chem 390:89–111

Part VI

Microfluidic Chips for Human Cell-Based Culture and Analysis

Chapter 13

Microfluidic 3D Cytotoxic Assay

Hyeri Choi, Sunghun Cheong, Ailian Jin, Dohyun Park, and Noo Li Jeon

Abstract

Microfluidic-based cytotoxic assays provide high physiological relevance with the potential to replace conventional animal experiments and two-dimensional (2D) assays. Here, a 3D method utilizing a microfluidic platform for analysis of lymphocyte cytotoxicity is introduced in detail, including platform design, cell culture method, real-time cytotoxic assay setup, and image-based analysis. A 2D experimental method is used for comparison, which effectively demonstrates the advantages of 3D microfluidic platforms in closely recapitulating immune responses within the tumor microenvironment. Moreover, a wide range of experimental possibilities and applications using microfluidic 3D cytotoxic assays is introduced in this chapter, along with their capabilities, limitations, and future outlook.

Key words Microfluidics, Organ-on-a-chip, Immunotherapy, Cytotoxic assay, Tumor microenvironment

1 Introduction

The development of immunotherapy gives us a new strategy to defeat cancer [1]. To advance the precision and efficacy of cancer immunotherapy, the mechanisms and functions of immune cells engaged in tumor progression should be clearly defined. This requires a deep understanding of the intricate interactions between the tumor microenvironment (TME) and the immune system [2–4]. Genetically engineered immune cells expressing chimeric antigen receptors (CAR), checkpoint inhibitors, and cytokines are typical immunotherapy approaches to enhance the cytotoxicity of immune cells and have shown remarkable clinical efficacy in treating hematopoietic cancers [5, 6]. However, the limited infiltration capacity of immune cells within solid tumors remains challenging due to the physical barriers created by the extracellular matrix (ECM) and the immunosuppressive environment [7]. The

The authors Hyeri Choi, Sunghun Cheong, and Ailian Jin have contributed equally to this chapter

Valérie Taly et al. (eds.), *Microfluidics Diagnostics: Methods and Protocols*, Methods in Molecular Biology, vol. 2804,
https://doi.org/10.1007/978-1-0716-3850-7_13,

development of physiologically relevant in vitro models capable of recapitulating the complex three-dimensional (3D) TME holds significant promise. Such models can serve as powerful tools for evaluating the infiltration and cytotoxic ability of immune cells in response to immunotherapy.

Animal models and simple 2D in vitro models have been widely used for assessing immune cell cytotoxicity. 2D cytotoxic assays conducted in multi-well plates offer a straightforward experimental setup and yield high-throughput results. Meanwhile, animal models, such as syngeneic mouse models, provide valuable insights into the complex interplay between the TME and the immune system, providing detailed information on tumor size progression and immune cell distributions within solid tumors. However, these conventional models possess inherent limitations, such as a lack of physiological complexity in recapitulating human organs, time- and cost-inefficiency associated with animal experimentation, and a reduced chance of successful translation to clinical trials due to genetic disparities across species [8–11].

In contrast, microfluidic 3D cytotoxic assays utilizing human cells offer distinctive advantages over conventional cytotoxic assays. These assays deliver high-throughput capabilities, embed more physiologically relevant 3D environments, and entail a cost-effective process with reduced reagent and cell consumption. They also enable precise control of physical and chemical aspects of human organ systems, such as hypoxic conditions, matrix stiffness, flow rate, and chemotactic gradients [12]. Furthermore, by integrating these microfluidic systems with standard microscopy imaging techniques, substantial steps of immune cell killing, from migration to recognition and lysis of target cells, can be monitored and analyzed in real time.

This chapter describes a proposed protocol using an injection-molded microfluidic platform and other recent studies on microfluidic 3D cytotoxic assays, with a particular focus on their platform designs, analysis methods, and the outcomes obtained from cytotoxic assays. It also discusses current limitations and the future outlook to promise more physiologically relevant in vitro models using microfluidic platforms.

Noo Li Jeon's research group at Seoul National University introduced an injection-molded microfluidic platform for the assessment of lymphocyte cytotoxicity within a 3D environment [13]. The platform comprises multiple wells for high-throughput assays and employs rail-based microstructures with injection holes to simplify hydrogel patterning through straightforward pipetting procedures [14]. In this platform, HeLa cells were embedded in collagen matrices to recapitulate a 3D environment and to allow natural killer (NK) cells to infiltrate and migrate toward cancer cells. The results show that the 3D ECM significantly reduces the cytotoxicity of NK cells compared to conventional 2D assays.

Interestingly, denser collagen matrices hindered migration and cytotoxic activity but promoted a faster lysis process compared to less dense collagen matrices. This model can evaluate the effect of physical properties of the 3D TME on the activity of cytotoxic lymphocytes and can be further used for testing immunotherapy to enhance their cytotoxicity.

2 Materials

2.1 Cell Culture

1. HeLa and NK-92 cells.
2. Cell culture medium: Dulbecco's modified eagle's medium (DMEM) mixed with 1% penicillin-streptomycin (PS) and 10% fetal bovine serum (FBS).
3. Minimum essential media: Minimum essential media alpha (MEM α) with 15% horse serum, 15% FBS, and 1% PS and other supplements such as 0.2 mM myo-inositol, 0.02 mM folic acid, and 0.1 mM 2-mercaptoethanol.
4. Interleukin-2.
5. T25 flasks.
6. 24 well plate.
7. Trypsin-EDTA: 0.25% trypsin-EDTA (HyClone).
8. CellTrace CFSE Cell Proliferation Kit (Thermo Fisher, C34570).
9. CellTrace Far Red Cell Proliferation Kit (Thermo Fisher, C34572).
10. Propidium iodide (PI).
11. Collagen gel pre-solution: 9 mg/mL of rat tail-oriented collagen type I (Corning, 354249) mixed with 150 mM HEPES buffer at a 2:1 ratio (v/v).

2.2 Equipment

1. Air plasma setup to treat the platform.
2. Cell culture incubator.
3. Centrifuge.
4. Nikon Eclipse Ti-E inverted microscope system.
5. Nikon Ti 2 A1 confocal microscope.

3 Methods

3.1 Platform Design

The injection-molded platform consists of rail-based structures, injection holes, and medium reservoirs in order to cultivate multicellular components in a 3D environment. Figure 1 shows a picture of the microfluidic platform with corresponding schematic top and

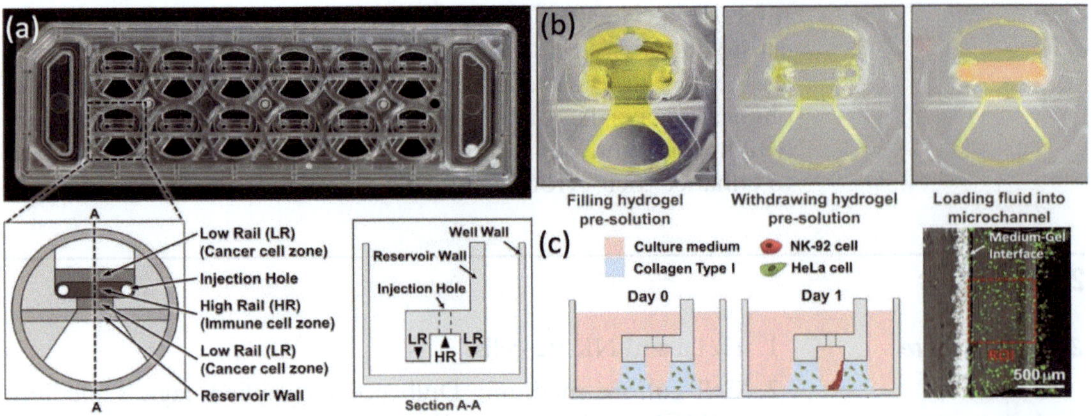

Fig. 1 Microfluidic platform for cytotoxic assays. (**a**) Visual representation of the platform with corresponding schematic top and cross-section view of an individual well. (**b**) Hydrogel patterning process under low rails (LR) and subsequent central high rail (HR) patterning (**c**) Schematic assay diagram: Loading HeLa cells in collagen under LR (Day 0), incubation, loading NK-92 cells under HR (Day 1) with subsequent tilting for NK cell accumulation and final incubation. Fluorescent image on the right shows HeLa cells (green) embedded in collagen and NK-92 cells (white) deposited on the medium-gel interface

cross-section views of an individual well within the platform. Once the microchannels with lower rails (indicated as cancer cell zones) are first filled with hydrogel pre-solution, immune cells in suspension are injected into the central channels (indicated as immune cell zone) through the injection hole. Detailed protocols for preparing the hydrogel pre-solution and conducting the cytotoxic assay are provided as follows.

3.2 Cell Culture

1. Culture HeLa cells using cell culture medium.
2. Culture NK-92 cells using minimum essential media.
3. Subculture NK-92 cells in T25 flasks every 2 days at a cell density of 10^5 cells/mL in 6 mL of minimum essential media with 1200 units of Interleukin-2 added.

3.3 3D Cytotoxicity Assay Using Gel Patterned Platform

1. Detach HeLa cells from the culture dish using Trypsin-EDTA. Obtain HeLa cell pellet through centrifugation.
2. Resuspend HeLa cells in serum-free cell culture medium with 2 μM of CellTrace CFSE Cell Proliferation Kit and incubate them for 30 min to label HeLa cells.
3. Mix the CFSE-labeled HeLa cells with collagen gel pre-solution.
4. Expose the platform to air plasma at 70 W for 3 min (*see* **Note 1**).
5. Inject the collagen pre-solution containing HeLa cells through the injection hole to pattern the area under the LR (*see* **Note 2**).

6. Incubate the platform in a cell culture incubator at 37 °C with 5% CO_2 for 30 min to induce crosslinking of the collagen.
7. Fill the two medium reservoirs with cell culture medium (*see* **Note 3**).
8. Incubate the platform in the cell culture incubator for another 24 h.
9. Obtain NK-92 cell pellets through centrifugation.
10. Resuspend NK-92 cells in serum-free minimum essential media with 2 μM of CellTrace Far Red Cell Proliferation Kit and incubate for 20 min to label NK-92 cells.
11. Aspirate culture medium from each well and load 2 μL of NK-92 cell suspension with a concentration of 2×10^6 cells/mL into the HR of the platform (*see* **Note 4**).
12. Incubate the platform in the incubator at an angle of 90° for 20 min to induce deposition of NK-92 cells on one of the collagen blocks that contains HeLa cells.
13. Fill the two medium reservoirs with minimum essential media (*see* **Note 5**). Add PI to the medium for live-cell imaging.
14. Monitor migration and cytotoxic activity of NK-92 cells for 24 h using a real-time live-cell imaging system.

3.4 Image Analysis

1. Acquire time-lapse images using an inverted microscope system.
2. Obtain endpoint images of NK-92/HeLa co-culture with a confocal microscope and perform optical z-sectioning with 100 μm depth and 4 μm interval (*see* **Note 6**).
3. Use auto threshold ("Mean" method) in Fiji to convert z-projected images into binary images and perform quantitative analysis.
4. Divide the total area of NK cells by the average single NK cell area to estimate the number of NK cells in each subregion (*see* **Note 7**).
5. Calculate the average area ratio single live and dead HeLa cells. Then, multiply the areas of dead HeLa cells by the calculated average area ratio to determine the areas of an equivalent number of live HeLa cells.
6. Calculate the percentage of killed HeLa cells by dividing the calculated dead HeLa cell area by the sum of the live and dead HeLa cell area.

3.5 2D Cytotoxicity Assay for Comparison

1. Load 6×10^4 cells/mL of CFSE-labeled HeLa cells in 1 mL of culture media into each well of a 24-well plate and culture for 24 h.

2. Aspirate culture media and add NK-92 cells to the wells at different E:T ratios and culture them for another 24 h.
3. Add 1 uM PI to the media.
4. Calculate the viability of HeLa cells by dividing the live HeLa cell area in the co-culture group by the live HeLa cell area in the control group without NK-92 cells.

3.6 2D Versus 3D Cytotoxic Assays

The limitations of 2D cell culture systems for drug screening, which lack the crucial features of the complex 3D TME, underscore the advantages offered by microfluidic platforms. Microfluidic 3D cytotoxic assays provide reliable and reproducible results by conducting high-throughput experiments and considering diverse environmental factors that influence cellular responses in the system.

The described protocol involved a comparative analysis between the results obtained from the microfluidic 3D cytotoxic assay and conventional 2D cytotoxic assay. Immune cells encounter numerous challenges in their journey through the bloodstream, traversing tight endothelial barriers and navigating toward target tumor sites. Conventional 2D cytotoxic assays lack consideration for these obstacles and merely observe the cytotoxic activity of NK cells against tumor cell monolayers. In contrast, microfluidic 3D cytotoxic assays involve complex interactions within a 3D immunosuppressive environment.

The proposed microfluidic 3D cytotoxic assay highlighted the significance of the 3D environment by considering the ECM surrounding tumor cells, which impedes lymphocyte access to tumor cells [13]. Figure 2 shows the real-time observation of NK cell cytotoxicity against HeLa cancer cells on the microfluidic platform. A comparative analysis revealed that the 2D cytotoxic assay exhibited a higher number of HeLa cell deaths induced by NK cells, suggesting that the microfluidic 3D assay can reflect the immunosuppressive effects resulting from the ECM. This microfluidic 3D cytotoxic assay holds promise as a potent tool for understanding immune-tumor cell interactions, evaluating emerging immunotherapies, and assessing their impacts on immune cell cytotoxicity and migration.

3.7 Replication of TME Components

Replicating essential components of the TME, including the immune system, vascular networks, stromal compartment, and immunosuppressive factors, is crucial for studying cancer immunotherapy in vitro. These components interact in complex ways that can either inhibit or promote tumor growth and immunosurveillance. For example, a hypoxic TME, characterized by limited oxygen supply, impairs immune responses against solid tumors. Ando et al. developed a tumor model using a polydimethylsiloxane (PDMS)-based chamber with an oxygen gradient to investigate

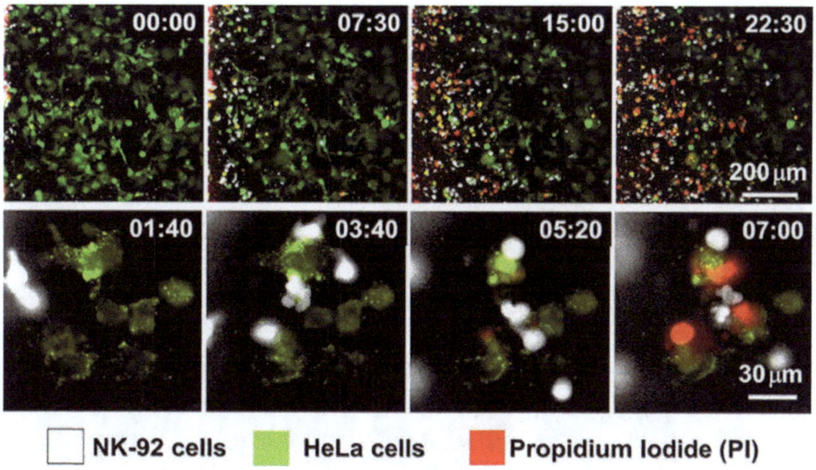

Fig. 2 Live imaging capturing the migration and cytotoxicity dynamics of NK-92 cells against HeLa cells within the collagen matrix. NK-92 and HeLa cells were labeled with white and green fluorescence, respectively. The time-lapse sequence illustrates NK-92 cells migrating within the collagen matrix and actively participating in the cytotoxic elimination of HeLa cells in their proximity, as evidenced by the increasing number of PI-stained HeLa cells

the impact of hypoxia on the cytotoxicity of CAR-T cells [15]. Under hypoxic conditions, several immunosuppressive mechanisms, such as upregulated PD-L1 expressions in tumor cells, were observed to reduce T cell-mediated cytotoxicity. CAR-T cells showed enhanced cytotoxicity and migratory capabilities compared to non-transduced T cells in the same hypoxic environment.

Vascularized tumor models using microfluidic platforms have been highly desired for replicating the TME, as blood vessels continuously supply nutrients and oxygen and serve as transportation routes for drugs and immune cells. The proposed protocol in this chapter did not include vascularized tumor models. To address this limitation, Song et al. from the same research group introduced an advanced model (Fig. 3a) that included self-assembled tumor vasculature [16]. The presence of firmly perfusable vessels within the platform enabled NK cells to effectively extravasate and eliminate target tumor cells. Intriguingly, NK cells demonstrated varying cytotoxic abilities depending on the subtypes of colorectal cancer cells in this model.

Using a vascularized tumor model, Pavesi et al. proposed an in vitro microfluidic cell extravasation assay to study complex cell-cell interactions within the TME. Their platform included a perfusable vascular network, mimicking the hepatic TME, and utilized engineered T cells designed to target hepatitis B virus (HBV) in hepatocellular carcinoma cells (HCC) to analyze T cell cytotoxicity [17]. Additionally, Nguyen et al. developed a tumor-on-chip platform that precisely reconstructed the TME with breast cancer cells,

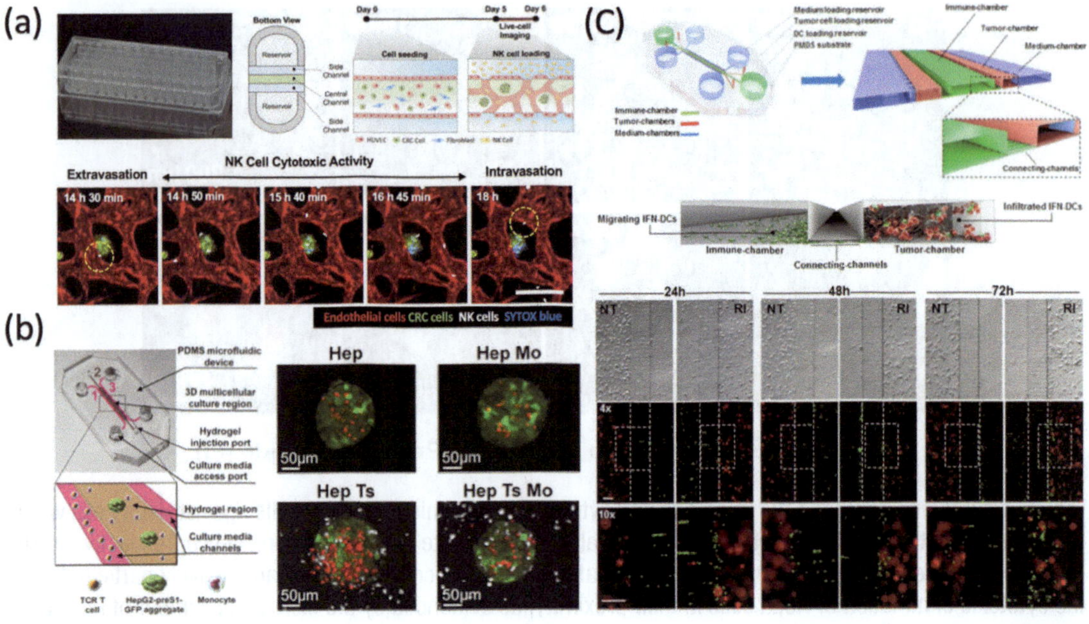

Fig. 3 Recent studies on microfluidic 3D cytotoxic assays. (**a**) Injection-molded 3D microfluidic platform showing two side channels, a central channel acting as fluid guide rails, and two reservoirs for growth medium. Upper right image indicates critical time points: cell seeding, NK cell loading, and live-cell imaging for a tumor vasculature model. Fluorescent images demonstrate the subsequent process of NK cell cytotoxic activity [16]. (**b**) 3D microfluidic platform showcasing diverse cell configurations to replicate a multicellular tumor microenvironment. Lower images demonstrate cytotoxicity exerted by HBV-specific TCR T cells (white) and dead target cells (red). A reduction in TCR T cell cytotoxicity is observed in the presence of monocytes [25]. (**c**) Schematic diagram of a 3D microfluidic platform featuring a central chamber for culturing DCs and two lateral tumor chambers. Fluorescent images highlight the enhanced migration of DCs toward the treated tumor chamber (indicated as RI) [31]

immune cells, endothelial cells, and cancer-associated fibroblasts [18]. Interestingly, cancer-associated fibroblasts specifically counteracted the inhibitory and cytotoxic effects of Trastuzumab, an antibody cancer drug targeting HER2 protein, on breast cancer cells.

3.8 Understanding Immune Cell Functions in the TME

Microfluidic platforms have emerged as a powerful tool offering a unique approach to investigate the interactions between various immune cell types [19–23]. Aung et al. developed a photopatterning-based microfluidic platform that contains tumor cells, monocytes, and endothelial cells to study immune-tumor cell interactions and their effects on T cell recruitment [24]. The study showed that cancer spheroids were more effective at recruiting T cells compared to dispersed cells or small cell aggregates. Additionally, THP-1 monocytes not only induced greater T cell recruitment but also exhibited increased T cell infiltration. To examine the impact of human primary monocytes on the killing ability of HBV-specific T cell receptor (TCR)-T cells, Lee et al. developed a

3D microfluidic platform that recapitulates the HBV-HCC TME (Fig. 3b) [25]. This model consists of a multicellular culture region where monocytes and transduced HepG2 cells are co-cultured in collagen gel and culture media channels. The study found that monocytes reduced retrovirally transduced TCR-T cell cytotoxicity by triggering PD-L1/PD-1 signaling but did not affect mRNA-electroporated TCR-T cell cytotoxicity. This result was only observed in the 3D microfluidic model. Aside from the effects of monocytes on T cell activity, 3D microfluidic platforms can also be utilized to investigate the interactions between immune cells and biomolecules or drugs [26, 27].

Microfluidic platforms can be applied to investigate the functions of various immune cell subsets within the TME [2, 28]. For example, Bi et al. devised a platform to study the impact of defined macrophage subsets on tumor behavior. The platform is designed to recapitulate key features of parental tumors and contains three adjacent and connected chambers [29]. Both upper and lower chambers are loaded with tumor cells, and the central chamber contains a dynamic, perfused, living microvascular network. Co-culturing human pancreatic or colorectal cancer cells with M1-polarized macrophages exhibited significant inhibition of tumor growth and tumor-induced angiogenesis. In contrast, M2-polarized macrophages secreted matrix metalloproteinase-7 and angiopoietin, promoting tumor migration into the vascularized chamber and enhancing tumor growth and angiogenesis.

3.9 Therapeutic Approaches Using Combination Therapy

Combination therapy has emerged as a promising therapeutic approach with improved clinical outcomes. Several studies have shown the beneficial effects of combining immune cells with other anti-cancer drugs. Ayuso et al. developed a microfluidic model to study the cytotoxic activity of NK cells within solid tumors [30]. The presence of tight endothelial barriers covering lumen channels resulted in delayed penetration of both antibodies and NK cells, and NK cells showed distinct chemotactic responses depending on their distance from the tumor spheroid. The combination of NK cells and antibodies targeting EpCAM induced antibody-dependent cellular cytotoxicity that led to higher overall cytotoxicity. NK cell infiltration and antibody diffusion as well as directional migration were quantified using image-based analysis.

Parlato et al. developed a microfluidic platform containing a central immune chamber, two tumor chambers, and connecting chambers in between to monitor the behavior of human dendritic cells (DCs) toward tumor cells (Fig. 3c) [31]. Interferon-alpha (IFN-α)-conditioned DCs were used in combination with IFN-α and the epigenetic drug, romidepsin. A significantly higher number of DCs infiltrated into the connecting channel toward treated tumor cells compared to untreated tumor cells. Higher levels of chemokines, particularly CXCL12, were found in the treated tumor cells, which suggested that the CXCR4/CXCL12 pathway dominantly contributed to DC recruitment to the tumor sites.

3.10 Limitations and Future Outlook

Although microfluidic-based cytotoxic assays have shown clinically meaningful outcomes and demonstrated their potential in drug screening, they still have certain limitations. The assay proposed in this chapter primarily focuses on the interaction between the ECM, immune cells, and tumor cells. However, other advanced in vitro models containing vascular networks would be considered more physiologically relevant models as they can provide additional barriers that influence immune cell migration and lead to a more accurate assessment of cytotoxicity [32, 33]. It is also important to acknowledge that the absence of certain cellular components or unconsidered factors in in vitro models can play crucial roles in tumor progression and immune responses within the TME. When designing specific organ models, careful selection of components and avoiding oversimplification are essential to ensure that critical factors are not ignored while the model closely reflects the complexities of the TME.

One of the challenges in the organ-on-a-chip field is to enhance the reliability of in vitro models [34]. As organ-on-a-chip technology is relatively emerging, it requires additional validation and comparison with well-designed animal studies or patient clinical trial information to demonstrate its clinical correlation. Due to the insufficient reliability of existing microfluidic platforms, they have yet been adopted as standards in clinical and pharmaceutical industries [35]. To address this limitation, the use of patient-derived cells with clinical data and integration with other technologies such as organoid development are possible approaches toward personalized treatment and may help reduce reliance on animal experiments in the future.

The materials and fabrication methods widely used in microfluidic platforms also have limitations [36]. Specifically, PDMS-based microfluidic platforms face challenges in accurately determining effective drug doses for in vitro testing, as they tend to absorb small molecules. Furthermore, labor-intensive fabrication processes hinder their practical use as drug-screening tools [37]. Despite efforts to address these challenges, building complex microchannels and structures in a high-throughput manner remains challenging. Although injection molding can produce microfluidic devices at scale, it has limitations in resolution. Advancements in 3D printing, along with the development of biocompatible and transparent resin with high resolution, enable the fabrication of complex structures with high accuracy, thus enhancing microfluidic 3D cell culture platforms.

Lastly, the wide variety of platform designs, materials, and formats make it difficult to apply microfluidic 3D culture models for drug screening purposes, as there are no clear standardized evaluation criteria for performance comparison between different platforms. The establishment of such standardized evaluation criteria is under active discussion among researchers, administrators, and pharmaceutical companies, but it will require significant time and efforts [38].

4 Notes

1. Use any commercially available Cell Proliferation Kit wavelength as needed. However, it is recommended to opt for a long wavelength for immune cell staining to minimize potential phototoxicity during live-cell imaging.
2. The power and time settings for air plasma treatment were optimized for the specific injection-molded microfluidic platform used in this assay. Adjustments may be necessary if a different platform is used.
3. After injecting the hydrogel with HeLa cells, gently aspirate the hydrogel again to precisely pattern the LRs.
4. Ensure that the injection hole of the high rail is also aspirated to prevent any retention of excess culture medium under the rail.
5. Carefully add minimum essential media to the medium reservoirs, considering NK-92 cells are suspension cells. Although there is sufficient time for cell accumulation on the side of the collagen, they may disperse due to forces generated during media addition.
6. Maintain consistent experimental conditions, including humidity, CO_2 levels, and temperature, between the live-cell imaging system and the incubation system.
7. For effective observation of NK cell migration and cytotoxic activity, the region of interest was divided into seven subregions, each with a width of 100 μm, based on the dimensions of the platform used in the assay. The first region was excluded from statistical analysis to mitigate the potential accumulation of NK cells.

References

1. Galon J, Bruni D (2020) Tumor immunology and tumor evolution: intertwined histories. Immunity 52:55–81
2. Zhang J et al (2022) Immunotherapy discovery on tumor organoid-on-a-chip platforms that recapitulate the tumor microenvironment. Adv Drug Deliv Rev 187:114365
3. Bader JE, Voss K, Rathmell JC (2020) Targeting metabolism to improve the tumor microenvironment for cancer immunotherapy. Mol Cell 78:1019–1033
4. Buoncervello M, Gabriele L, Toschi E (2019) The Janus face of tumor microenvironment targeted by immunotherapy. Int Journal of Mol Sci 20:4320
5. Khalil DN et al (2016) The future of cancer treatment: immunomodulation, CARs and

combination immunotherapy. Nat Rev Clin Oncol 13:273–290
6. Sterner RC, Sterner RM (2021) CAR-T cell therapy: current limitations and potential strategies. Blood Cancer J 11:69
7. Giraldo NA et al (2019) The clinical role of the TME in solid cancer. Br J Cancer 120:45–53
8. Rongvaux A et al (2014) Development and function of human innate immune cells in a humanized mouse model. Nat Biotechnol 32: 364–372
9. Vesely MD et al (2011) Natural innate and adaptive immunity to cancer. Annu Rev Immunol 29:235–271
10. Tseng D et al (2013) Anti-CD47 antibody mediated phagocytosis of cancer by macrophages primes an effective antitumor T-cell response. Proc of Nat Acad of Sci 110: 11103–11108
11. Hirt C et al (2014) In vitro 3D models of tumor-immune system interaction. Adv Drug Deliv Rev 79-80:145–154
12. Sontheimer-Phelps A, Hassell BA, Ingber DE (2019) Modelling cancer in microfluidic human organs-on-chips. Nat Rev Cancer 19(2):65–81
13. Park D et al (2019) High-throughput microfluidic 3D cytotoxicity assay for cancer immunotherapy (CACI-IMPACT platform). Front Immunol 10:1133
14. Park D et al (2021) Aspiration-mediated hydrogel micropatterning using rail-based open microfluidic devices for high-throughput 3D cell culture. Sci Rep 11:19986
15. Ronteix G et al (2022) High resolution microfluidic assay and probabilistic modeling reveal cooperation between T cells in tumor killing. Nat Commun 13:3111
16. Song J et al (2021) High-throughput 3D in vitro tumor vasculature model for real-time monitoring of immune cell infiltration and cytotoxicity. Front Immunol 12:733317
17. Pavesi A et al (2015) Using microfluidics to investigate tumor cell extravasation and T-cell immunotherapies. Annu Int Conf IEEE Eng Med Biol Soc 2015:1853–1856
18. Nguyen M et al (2018) Dissecting effects of anti-cancer drugs and cancer-associated fibroblasts by on-chip reconstitution of immunocompetent tumor microenvironments. Cell Rep 25:3884–3893.e3
19. Businaro L et al (2013) Cross talk between cancer and immune cells: exploring complex dynamics in a microfluidic environment. Lab Chip 13:229–239
20. Beckwith AL, Velásquez-García LF, Borenstein JT (2019) Microfluidic model for evaluation of immune checkpoint inhibitors in human tumors. Adv Healthc Mater 8:e1900289
21. Cui X et al (2020) Dissecting the immunosuppressive tumor microenvironments in glioblastoma- on-a-chip for optimized PD-1 immunotherapy. elife 9:e52253
22. Fang T et al (2019) Remodeling of the tumor microenvironment by a chemokine/anti-PD-L1 nanobody fusion protein. Mol Pharm 16: 2838–2844
23. Biselli E et al (2017) Organs on chip approach: a tool to evaluate cancer-immune cells interactions. Sci Rep 7:1–12
24. Aung A et al (2020) An engineered tumor-on-a-chip device with breast cancer-immune cell interactions for assessing T-cell recruitment. Cancer Res 80:263–275
25. Lee SWL et al (2018) Characterizing the role of monocytes in T cell cancer immunotherapy using a 3D microfluidic model. Front Immunol 9:416
26. Ayuso JM et al (2021) Microfluidic tumor-on-a-chip model to evaluate the role of tumor environmental stress on NK cell exhaustion, science. Advances 7:eabc2331
27. Ke L-Y et al (2018) Cancer immunotherapy μ-environment LabChip: taking advantage of optoelectronic tweezers. Lab Chip 18:106–114
28. Surendran V et al (2021) A novel tumor-immune microenvironment (TIME)-on-chip mimics three dimensional neutrophil-tumor dynamics and neutrophil extracellular traps (NETs)-mediated collective tumor invasion. Biofabrication 13:035029
29. Bi Y et al (2020) Tumor-on-a-chip platform to interrogate the role of macrophages in tumor progression. Integr Biol (Camb) 12:221–232
30. Ayuso JM et al (2019) Evaluating natural killer cell cytotoxicity against solid tumors using a microfluidic model. Onco Targets Ther 8: 1553477
31. Parlato S et al (2017) 3D microfluidic model for evaluating immunotherapy efficacy by tracking dendritic cell behaviour toward tumor cells. Sci Rep 7:1–16
32. Parlato S et al (2021) Tumor-on-a-chip platforms to study cancer–immune system crosstalk in the era of immunotherapy. Lab Chip 21:234–253
33. Kim S et al (2021) Microfluidic tumor vasculature model to recapitulate an endothelial immune barrier expressing FasL. ACS Biomater Sci Eng 7:1230–1241

34. Miccoli B, Braeken D, Li Y-CE (2018) Brain-on-a-chip devices for drug screening and disease modeling applications. Curr Pharm Des 24:5419–5436
35. Iyer V et al (2022) Advancing microfluidic diagnostic chips for clinical use. Lab Chip 22: 3110–3121
36. Ren K, Zhou J, Wu H (2013) Materials for microfluidic chip fabrication. Acc Chem Res 46:2396–2406
37. Van Meer B et al (2017) Small molecule absorption by PDMS in the context of drug response bioassays. Biochem Biophys Research Comm 482:323–328
38. Leung CM et al (2022) A guide to the organ-on-a-chip. Nat Rev Methods Primers 2:33

Chapter 14

Evaluating Nanoparticles Penetration by a New Microfluidic Hydrogel-Based Approach

Saba Goodarzi, François Lux, and Charlotte Rivière

Abstract

Reliable predictions for the route and accumulation of nanotherapeutics in vivo are limited by the huge gap between the 2D in vitro assays used for drug screening and the 3D physiological in vivo environment. While developing a standard 3D in vitro model for screening nanotherapeutics remains challenging, multi-cellular tumor spheroids (MCTS) are a promising in vitro model for such screening. Here, we present a straightforward and flexible 3D-model microsystem made out of agarose-based micro-wells, which enables the formation of hundreds of reproducible spheroids in a single pipetting. Immunostaining and fluorescent imaging, including live high-resolution optical microscopy, can be done in situ without manipulating spheroids.

Key words Hydrogel-based microsystem, Nanoparticles, Multicellular tumor spheroids (MCTS), High content optical imaging, 3D in vitro model

1 Introduction

Despite an increasing number of innovative therapeutic strategies, the majority of drug candidates, including nanotherapeutics, fail in clinical trials or in late preclinical tests [1]. This failure is largely due to the difficulty of modeling and predicting the transport of these nanoparticles in vivo [2] and their therapeutic efficacy, especially on the target site.

Predictive in vitro assessment of tissue penetration capacity remains one of the most difficult aspects of preclinical testing for any new nanomedicine [3]. Such information cannot be obtained from 2D cell cultures. Indeed, in conventional uptake studies, drugs and nanomedicines are added on top of cancer cells grown on flat monolayers, with the cell membrane serving as the only

Supplementary Information The online version contains supplementary material available at https://doi.org/10.1007/978-1-0716-3850-7_14.

Valérie Taly et al. (eds.), *Microfluidics Diagnostics: Methods and Protocols*, Methods in Molecular Biology, vol. 2804,
https://doi.org/10.1007/978-1-0716-3850-7_14,

barrier to the intracellular compartment. In contrast, nanotherapeutics encounter different physiological barriers in vivo limiting their transport into the tumor tissue.

Many different in vitro approaches have been developed in the last decade to overcome the limitations of 2D assays. Among them, spheroids have been presented as predictive models to analyze in vitro the effect of nanotherapeutics in avascular tumors [4], by recapitulating cell-cell and cell-matrix interactions, as well as gradients of oxygen, nutrient, and metabolic wastes [5]. A large number of studies have reported that cellular responses to drug treatments in 3D cultures are likely to give better in vivo predictions compared to 2D cultures [6, 7]. Cells in such 3D structures exhibit higher resistance to chemotherapeutics [8], and radiotherapy [9], closer to the one found in vivo, possibly a result of the limited drug diffusion through the tumor cell mass and hypoxia [10, 11].

There are varieties of spheroid construction processes available today, each with its advantages and disadvantages [12], several of which have been improved for large-scale production under extremely repeatable conditions [13]. They primarily include the application of cell attachment-resistant surfaces or physical pressures to stimulate cell-to-cell contacts and encourage the creation of 3D spheroid structures [14]. However, in these approaches, spheroids are often transferred to another platform for functional characterization and drug screening, which is laborious and may impair spheroids' quality.

To increase handling and control, spheroids and tumoroids have then been integrated into a miniaturized in vitro format [15] and are now considered a relevant alternative platform for drug screening [16]. In line with these new approaches, we have developed agarose-based microsystems (Fig. 1) [17–21]. Agarose presents several advantages and is widely used in tissue engineering, while still under-used in microfluidic set-ups. The low cost and wide availability of agarose are appealing for large-scale production. Its tunable mechanical properties can reproduce the in vivo microenvironment stiffness and its porous nature enables the free diffusion of small to medium-sized molecules (size <30 nm) [22], assuring the supply of nutrients.

We will present in this chapter an easy-to-use process to fabricate agarose micro-wells enabling the formation of hundreds of reproducible spheroids with homogeneous shape and size, in a single pipetting event and compatible with the multi-well plate format classically used in cell biology (Fig. 1). It provides a user-friendly platform for in toto immunostaining and optical imaging, without having to physically cut the spheroids.

Thanks to these agarose-based microwells, all steps of experiments from cell seeding, treatment, and fixation, to immunostaining could be performed in classical multi-well plates without transferring spheroids. However, in case of requiring transfer of

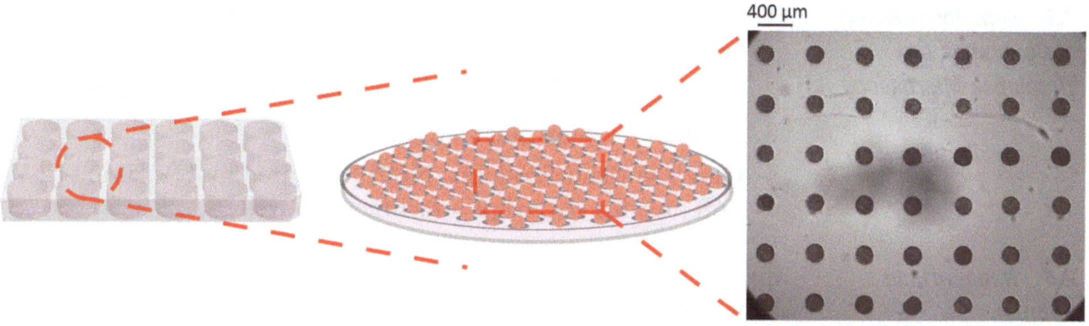

Fig. 1 Placement of agarose microwells in a 24-well plate, schematic of spheroids in agarose microwells (cylindrical, $d = 200$ μm), and image of several homogenous spheroids via optical microscopy (130 spheroids in each well and more 3000 spheroids in each 24-well plate compared to 96 spheroids in each ultra-low adhesion (ULA) 96-well plate)

spheroids, the microwells can be transferred easily and without manipulating spheroids. Another advantage of these microwells is their compatibility with high-throughput optical microscopy techniques, biological assays, or other analysis techniques (using either intact spheroids or single cells after spheroid dissociation).

2 Materials

2.1 Mold Preparation

1. Silicon wafer master mold: The desired design (number of wells, distance between wells, width, and depth of the wells) can be chosen and drawn using classical software used for photolithography such as CleWin4. Classical photolithography techniques will enable to get cylindrical wells. As an example, in this chapter, the master mold used contains 130 cylindrical microwells with a diameter and height of 200 μm. To get molds with a curved bottom, it is also possible to use micromilling [21] or 3D printing for the preparation of the master mold. The master molds are made with the same polarity as the final agarose molds (i.e., micro-wells), while PDMS, as a replica mold, will be with inverse polarity (i.e., with micro-pillars, see below).
2. Polydimethylsiloxane (PDMS) replica mold: PDMS replica molds are created by casting PDMS on the master mold. As many replica molds as needed can be made and will be used routinely to make the agarose-based microwells.

2.2 Agarose Solution

1. Agarose solution: 2% agarose solution (1 gr ultrapure agarose (Invitrogen) powder in 50 mL deionized water, *see* **Note 1**).

2.3 Functionalization of Coverslips for Timelapse Microscopy

1. Functionalization Reagents: 3-aminopropyltriethoxisilane (APTS) (Acros), acetic acid (vwr).
2. 12-well plate with drilled wells (holes diameter: 16 mm).
3. Round coverslips (diameter: 20 mm).
4. Glue (Norland Products, NOA 81) cured by UV lamp (12 W, 365 nm) for coverslips attachment to wells.

2.4 Nanoparticle Solution

Any nanoparticle solution can be assessed with our technique, as far as they are fluorescently labeled. As an example, AGuIX® nanoparticles are detailed below. In addition to AGuIX® labeled with Cy 5.5, fluorescent polymer nanoparticles and polymer chain penetration (20–100 nm in size) within spheroids have been studied using our hydrogel-based microsystem.

1. AGuIX® nanoparticles: The Gd-based nanoparticles (AGuIX®) synthesized by NH TherAguix (Lyon, France) are composed of a polysiloxane matrix surrounded by covalently bound DOTAGA-Gd ((1,4,7,10-tetraazacyclodode-cane-1-glutaric acid-4,7,10-triacetic acid)-Gd).

   The chemical composition of AGuIX® nanoparticles is ($GdSi_{6.5}N_6C_{25}O_{21}H_{42}$, 10 H_2O)n with a molar mass around 10–20 kDa. The hydrodynamic diameter of the AGuIX® nanoparticles is close to 5 nm, and the AGuIX® nanoparticles are characterized by a zeta potential of 9.0 ± 5.5 mV at pH 7.2 [23]. In the manuscript, all the concentrations concerning AGuIX® nanoparticles are expressed in terms of gadolinium concentration.

   These AGuIX® nanoparticles are further conjugated to Cyanine-5.5(Cy5.5) fluorophore to make them detectable by confocal fluorescence microscopy.
2. Mother nanoparticles solution: AGuIX®-Cy5.5 nanoparticles are reconstituted at a concentration of 100 mM of Gd in distilled water.
3. Nanoparticles solution for the incubation with spheroids: AGuIX®-Cy5.5 solutions are prepared with fresh full culture medium (DMEM, 10% FBS, 1% Penicilin-Streptomycine) to get a final Gd concentration of 0.8, 1.5 and 2 mM.

2.5 Solution for Immunostaining Procedures

1. Fixation solution: 4% PFA in PBS (stock solution at 16%, from Electron Microscopy Science, ref. 15710, diluted at 4% as working aliquots and stored at −20 °C).
2. Permeabilization solution: 0.1% Triton 100× solution in PBS (v/v).
3. Blocking solution: 3% BSA (w/v), 0.3% Triton 100× (v/v) in PBS (*see* **Note 2**).

4. Primary, secondary, fluo-marker solution: 1% BSA (w/v),0.3% Triton 100× (v/v) in PBS.
5. Rinsing solution: 1% BSA (w/v), 0.3% Triton 100× (v/v) in PBS.
6. Clarification solution (glycerol solution): 80% Glycerol in PBS (v/v). (to be prepared on the day of the experiment from a pure glycerol solution).

2.6 Nuclei and Organelle Staining

1. Nuclei staining: nucgreen™-Dead 488 (Invitrogen™) 1 drop/5 mL in PBS at 4 °C overnight.
2. Actins staining: phalloidin-546 solution (Invitrogen™) 1:50 in PBS at 4 °C overnight.
3. Early endosome staining: EEA1 (CellSignaling Technology), dilution: 1:100 in 1% BSA (w/v), 0.3% Triton 100× (v/v) PBS solution, at 4 °C overnight.
4. Mitochondria staining: AIF (CellSignaling Technology), dilution: 1:400 in 1% BSA (w/v), 0.3% Triton 100× (v/v) PBS solution, at 4 °C overnight.
5. Lysosome staining: LAMP-1 (CellSignaling Technology), dilution: 1:200 in 1% BSA (w/v), 0.3% Triton 100× (v/v) PBS solution, at 4 °C overnight.
6. Secondary antibody: Goat-anti rabbit IgG-Alexa 555 (Invitrogen™, A21428) dilution: 1:500 in 1% BSA (w/v), 0.3% Triton 100× (v/v) PBS solution, at 4 °C overnight.

2.7 Materials for Optical Microscopy Mouting

1. Coverslips 22 × 22 mm/76 mm × 26 mm.
2. iSpacer® ×2 (22 × 30mm, 0.5 mm thickness, Sunjinlab, containing 4 holes of 7 mm in diameter).
3. Thin double-sided sticky iSpacer (0.05 mm).
4. 80% Glycerol in PBS (v/v).
5. Glass Microscope slides 76 × 26 mm.
6. 6 mm puncher (Biopunch®, healthyLink), or yellow tip.
7. Confocal microscope.

3 Methods

3.1 Agarose Microwells Preparation

1. Autoclave agarose and water at 121 °C are used to dissolve agarose (*see* **Note 1**). A video is available to visualize the following steps (Video 1).
2. Prewarm PDMS and tips (with the edge cut) on the hot plate at 78 °C.
3. Put the liquid standard agarose on the hot plate at 100 °C.

4. Put 130 μL of agarose on the PDMS mold on the hot plate.
5. Put a coverslip on top and remove the PDMS+coverslip from the hot plate.
6. Wait for 10 min for the agarose to gel.
7. Gently unmold the gel by gently pushing on the PDMS mold. The agarose mold will remain stuck on the coverslip.
8. Remove the molded agarose from the coverslip and put it in a 12- or 24-well plate.
9. Add about 1 mL PBS to each well containing a system.

3.2 Agarose Microwells on APTS-Functionalized Coverslips

For time-lapse microscopy, the agarose molding needs to be performed on the APTS-functionalized coverslips to maintain the micro-wells (x,y,z) position over time. The preparation steps are described below:

1. Drill holes in each well of a 12-well plate (diameter 16 mm) to prepare the plate for the attachment of coverslips.
2. Prepare a functionalization solution with 1% APTS–5 mM acetic acid in distilled water (12.5 μL of acetic acid in 43.5 mL distilled water. Homogenize the solution. Add 435 μL of 3-APTS).
3. Incubate round coverslips (diameter 20 mm) in APTS solution for 20 min under stirring conditions.
4. Rinse coverslips extensively with water and dry on a hot plate (100 °C, 15 min).
5. Use the APTS-functionalized coverslips immediately for agarose molding using the same procedure as the one described above for free-standing microwells.
6. After agarose gelation, remove the PDMS mold; the agarose microwells remain attached to the APTS-functionalized coverslip.
7. Glued these coverslips with microwells on the bottom of the 12-well plate using curing optical adhesive activated by 30 s of exposure to a UV lamp (12 W, 365 nm).

3.3 UV Sterilization

1. *UV sterilization*: if possible use a dedicated UV lamp (8 W, 254 nm) to sterilize the multi-well plate containing the agarose molds (*see* **Note 3**): 20 min minimum without lid/20 min with lid (Fig. 2a).
2. The sterilized multi-well plate can be stored at 4 °C for a few days before use (put parafilm around to avoid evaporation).

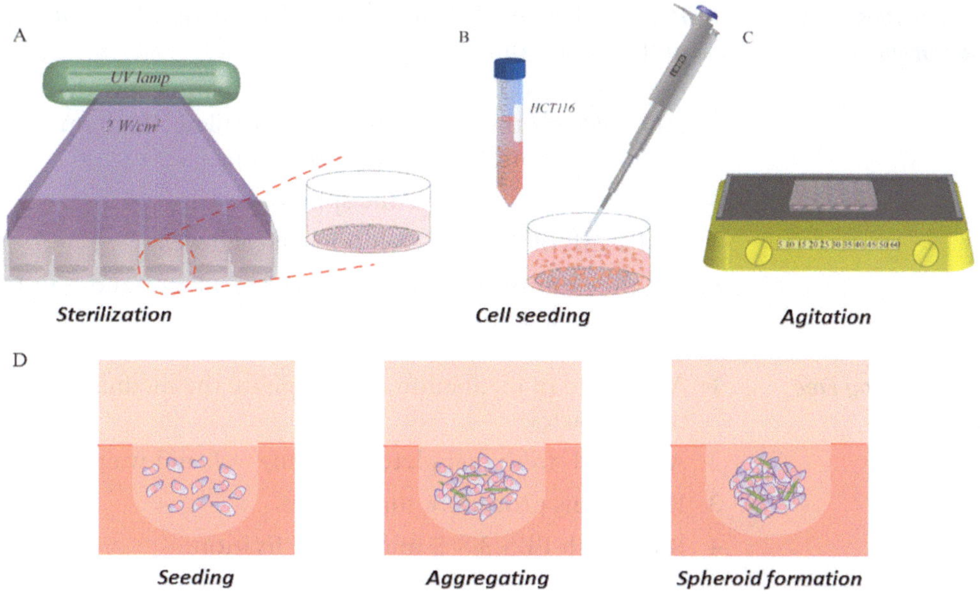

Fig. 2 Procedure of spheroids preparation using agarose microwells, (**a**) sterilization with UV lamp, (**b**) cell seeding, (**c**) agitation, and (**d**) steps of spheroids formation via self-assembly in microwells

3.4 Cell Seeding

1. The day before cell seeding, remove PBS in all wells. Replace by complete culture medium (1 mL per well for a 24 multi-well plate). Put the plate in an incubator at 37 °C and 5% CO_2 overnight.
2. After overnight incubation, detach the cells from your culture flasks according to standard protocols.
3. After centrifugation, prepare a cell suspension of 1.2 10^5 cells/mL.
4. Remove the complete culture medium from each multi-well (each containing 1 microsystem).
5. Add 1 mL of cell suspension at 1.2 10^5 cells/mL (Fig. 2b).
6. To encourage and accelerate cell aggregation, the 24-well plate is placed under orbital agitation (160 rpm) for 15 min in an incubator at 37 °C and 5% CO_2 (Fig. 2c).
7. After 4 h at 37 °C without agitation, rinse the plate with fresh medium to remove cells that did not reach the microwells.
8. Add 1 mL fresh complete medium/well.
9. Put the multi-well plate back into the incubator and wait for cell self-aggregation (2–3 days depending on cell type) (Fig. 2d).

3.5 Incubation with Nanoparticles

1. Prepare 100 mM AGuIX® solution in distilled water as the mother solution.
2. Prepare AGuIX® solutions for incubation by diluting the 100 mM mother solution in a complete medium to make 0.8 mM, 1.5 mM, and 2 mM solutions.
3. Aspirate medium above the spheroids.
4. Add 1 mL of AGuIX® solutions in each well.
5. Put the multi-well plate back in the incubator and wait for the desired incubation time.

3.6 Eliminating Free Nanoparticles

1. At the end of incubation time, aspirate the medium containing nanoparticles.
2. Add 1 mL fresh complete medium/well and wait for 15 min.
3. Repeat this procedure three times.
4. Rinse with PBS 3 × 5 min before fixation.

3.7 Fixation and Immunostaining

The developed 3D in vitro model in this study is compatible with in situ immunostaining, which is one of its main advantages. Indeed, this 3D in vitro model allows all steps of these types of experiments, from cell seeding to fixation and immunostaining, to be performed in the same plate without transferring spheroids. This makes it possible to have multiple labeling in several spheroids in one experiment, which is especially useful to find the localization of nanoparticles inside spheroids and within cells and subsequent quantification of these images.

1. Remove PBS in all wells. Replace with a 4% PFA solution (250 μL/well for a 24-well plate). Incubate 20 min at Room Temperature. Rinse with PBS 3 × 5 min.
2. Remove PBS in all wells. Replace with the permeabilization solution (500 μL/well for a 24-well plate). Incubate 10 min at Room Temperature.
3. Remove the permeabilization solution. Replace with the blocking solution. Incubate 1 h at room temperature.
4. Remove the blocking solution. Incubate with the primary antibody overnight at 4 °C (200 μL/well for a 24-well plate).
5. Remove the primary antibody solution. Rinse 3 × 5 min with the rinsing solution. Incubate with the secondary antibody overnight at 4 °C.

3.8 Clarification

Optical sectioning offers a potentially fast, simple, and low-cost alternative for the three-dimensional reconstruction of fluorescently labeled structures at subcellular resolution. Tissue opacity and light scattering, however, limit the utility of optical sectioning for deep imaging. An optimal clarification technique is necessary for clearing the spheroids and optimizing confocal fluorescence microscopy. For small spheroids (up to 300 μm in diameter), efficient clarification is obtained using glycerol as a clarifying agent (*see* **Note 4**).

1. Remove the secondary antibody solution.
2. Rinse 3 × 5 min with PBS.
3. Incubate with the glycerol solution overnight at 4 °C.
4. Remove the solution and change with a fresh one.

3.9 Mounting Agarose Micro-wells on Coverslips for Optical Microscopy

1. Glue together 2 thick iSpacers (0.5 mm from Sunjin Lab), thanks to the thin double-sided sticky spacer (0.05 mm from Sunjin Lab).
2. Glue the iSpacers to the already cleaned glass slide coverslip.
3. Punch the agarose microwell gels with a 6 mm puncher (or a yellow tip).
4. Put the agarose microwells in the specific holes of the iSpacer.
5. Add 35 μL of a solution of 80% glycerol in PBS.
6. On top of the iSpacer, glue the clean coverslip (0.17 mm) using the iSpacer sticky part. The spheroids are now in contact with the thin coverslip.
7. Seal everything with transparent varnish. The samples are now ready for imaging (*see* **Note 5**).

3.10 Confocal Acquisition

Reversing microwells containing spheroids during image acquisition is used to bring the spheroids closer to the objective (from Fig. 3a to Fig. 3b). The spheroids are imaged in the z-direction, from the starting point in Fig. 3c on the outer side toward the center of the spheroid, taking images in each 1-micrometer thickness as optical sectioning (Fig. 3c, d), which provides a sequence of images at different depths of spheroids (Fig. 3d).

3.11 Quantification of Nanoparticle Penetration Within Spheroids

Images can be processed using ImageJ and Matlab routinely to quantify the fluorescent signal intensity (coming from the nanoparticles), as a function of the distance from the center of the spheroid (*see* **Note 6**). Matlab code can be provided upon request. First, the nucgreen™ channel is used to segment the area of spheroids in each slice (Fig. 4a), and these segmented regions are then fitted into a perfect circle (Fig. 4b), with all of these circles being used to estimate the spherical shape and radius of the spheroids (Fig. 4c). The results of this analysis are used to switch the direction of analysis from Cartesian coordinates (XYZ) to Spherical coordinates (r, θ, Φ) (Fig. 4d), and the mean intensity can be determined by averaging the intensity along angles θ and Φ, giving the mean intensity in certain radius in each spheroid (Fig. 4e). As a result,

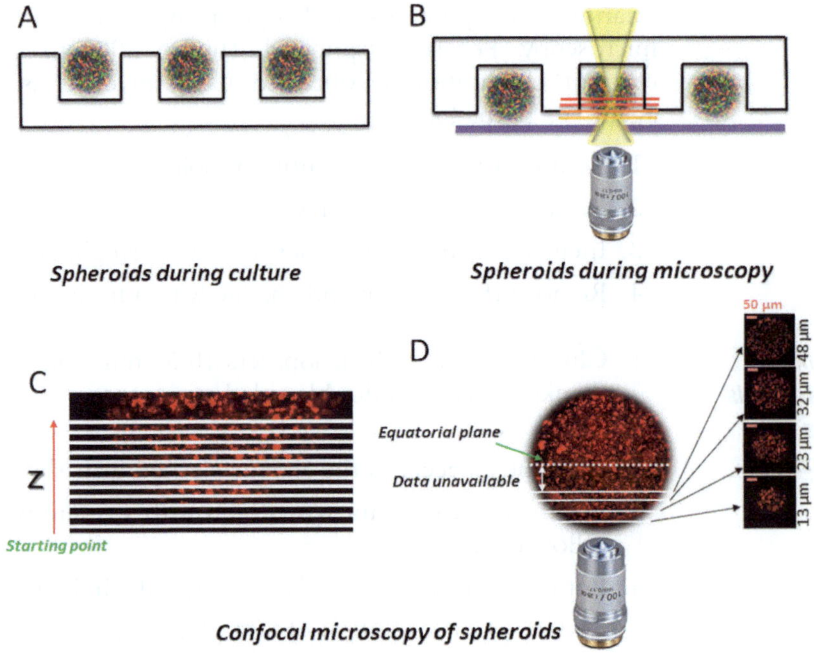

Fig. 3 Principle of confocal microscopy of spheroids in microwells, (**a**) placement of spheroids in microwells during culture and other steps, (**b**) placement of spheroids in agarose microwells regarding the objective during confocal microscopy, (**c**) direction of image acquisition in confocal microscopy, and (**d**) view of images taken by optical sectioning and limit of imaging deeper slices of spheroids

the mean fluorescence signal intensity could be calculated as a function of distance from the periphery of spheroids (Fig. 4f).

3.12 Quantification of Colocalization of Nanoparticles with Subcellular Organelles

To determine the degree of nanoparticle colocalization with cellular organelles, images can be analyzed using a routine developed in Matlab to calculate Pearson's Correlation Coefficient as a statistical and quantitative indicator for colocalization (*see* **Note 7**). Matlab code can be provided upon request.

First, in each slice, the entire spheroid is segmented using the nucgreen™ channel. Using this mask to outline the region of interest, the corr2 function in Matlab is used to calculate the Pearson Correlation Coefficient of two fluorophores belonging to AGuIX® and a cellular organelle.

3.13 ICP Analysis

In parallel, concentrations of Gd can be analyzed using inductively coupled plasma-mass spectrometry (ICP-MS). To prepare samples for this analysis, after spheroids incubation with nanoparticles, the spheroids are dissociated using the following steps.

1. Rinse spheroids with 1 mL PBS/well three times, each 15 min.
2. Dissociate spheroids by pipetting 1 mL trypsin + EDTA over spheroids.

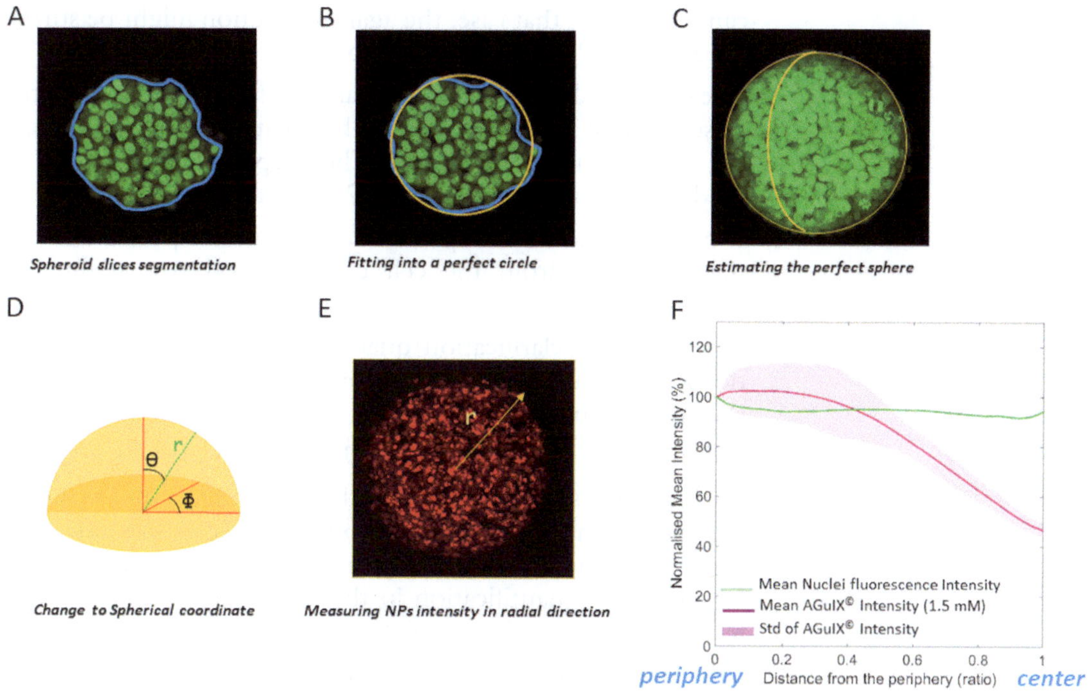

Fig. 4 Quantification of confocal images; (**a**) spheroid segmentation in one slice of HCT-116 spheroids labeled with Nucgreen™ -488 imaged via confocal microscopy. (**b**) fitting a segmented slice into a perfect circle. (**c**) Estimating spheroids slices in a perfect sphere and measuring their diameter. (**d**) Schematics of spherical coordinates in a spheroid. (**e**) z-projection of an imaged of spheroid incubated with AGuIX®—Cy5.5 showing the radius of the spheroid. (**f**) Curved obtained via image analysis in spherical coordinates showing AGuIX®—Cy5.5 intensity in radial direction

3. Transfer them to 15 mL falcon.
4. Count cells.
5. Centrifuge dissociated spheroids (900 g for 5 min).
6. Discard the supernatant.
7. Dissolve cell pellets in 150 μL HNO_3 69% (ROTH) at 80 ° C for 3 h.
8. Add ultrapure water to increase the volume to 10 mL.
9. Measure Gd concentration with PerkinElmer, NexION® 2000 IC-MS.
10. Calculate the average Gd concentration per cell.

4 Notes

1. The agarose is preferably dissolved using an autoclave (20 min, 121 °C) to avoid bubble formation. If unavailable, micro-wave

can be used. In that case, the agarose solution might be stirred for 2 h at 90 °C to get rid of bubbles.

2. The solution should be prepared ahead of the experiment as the dissolution of BSA requires a few hours at 37 °C. Avoid vortexing to avoid foam formation. The prepared solution can be used for a few weeks if they are 0.2 μm-filtered and kept at 4 °C.
3. If no dedicated UV lamp is available, it is also possible to use the UV lamp from the cell culture hood (overnight UV exposure).
4. Assessment of clarification quality, and subsequent convolutional neural network-based nuclei segmentation is thoroughly investigated in [18].
5. Mounted gels can be kept at −20 °C for long-term storage.
6. Indeed, one should be aware that an image from a single z-plane cells has varying access to nanoparticles and nutrients. As a result, since it combines information from non-equal parts in spheroids, quantification in the z-direction is not accurate for 3D spherical cellular structures.
7. The most common method for analyzing colocalization in biological images is superposition of fluorescence images. The results, however, can be misleading because the intermediate color, which indicates colocalization, is only achieved if the intensities of two probes in fluorescence images are similar [24].

Acknowledgments

Hélène Delanoë-Ayari for her help in developing the Matlab code to quantify nanoparticle penetration within spheroids. Arnaud Favier for providing polymer nanoparticles to test their penetration in spheroids using our 3D in vitro model. Remy Fulcrand for his help in photolithography. Fabien Rossetti for his help in ICP analysis. Gilles Simon for his help in the preparation of the plate for time-lapse microscopy.

References

1. Dowden H, Munro J (2019) Trends in clinical success rates and therapeutic focus. Nat Rev Drug Discov 18(7):495–496
2. Wilhelm S et al (2016) Analysis of nanoparticle delivery to tumours. Nat Rev Mater 1(5): 16014
3. Minchinton AI, Tannock IF (2006) Drug penetration in solid tumours. Nat Rev Cancer 6(8):583–592
4. Millard M, Yakavets I, Zorin V, Kulmukhamedova A, Marchal S, Bezdetnaya L (2017) Drug delivery to solid tumors: the predictive value of the multicellular tumor spheroid model for nanomedicine screening. Int J Nanomedicine 12:7993–8007
5. Hirschhaeuser F, Menne H, Dittfeld C, West J, Mueller-Klieser W, Kunz-Schughart LA (2010) Multicellular tumor spheroids: an

underestimated tool is catching up again. J Biotechnol 148(1):3–15
6. Imamura Y et al (2015) Comparison of 2D- and 3D-culture models as drug-testing platforms in breast cancer. Oncol Rep 33(4): 1837–1843
7. Fontoura JC et al (2020) Comparison of 2D and 3D cell culture models for cell growth, gene expression and drug resistance. Mater Sci Eng C Mater Biol Appl 107:110264
8. Virgone-Carlotta A et al (2017) In-depth phenotypic characterization of multicellular tumor spheroids: effects of 5-fluorouracil. PLoS One 12(11):e0188100
9. Koch J et al (2021) Three dimensional cultivation increases chemo- and radioresistance of colorectal cancer cell lines. PLoS One 16(1): e0244513
10. Breslin S, O'Driscoll L (2016) The relevance of using 3D cell cultures, in addition to 2D monolayer cultures, when evaluating breast cancer drug sensitivity and resistance. Oncotarget 7(29):45745–45756
11. Fernandes S, Cassani M, Pagliari S, Filipensky P, Cavalieri F, Forte G (2020) Tumor in 3D: in vitro complex cellular models to improve nanodrugs cancer therapy. Curr Med Chem 27(42):7234–7255
12. Lv D, Hu Z, Lu L, Lu H, Xu X (2017) Three-dimensional cell culture: a powerful tool in tumor research and drug discovery (review). Oncol Lett 14(6):6999–7010
13. Costa EC, Moreira AF, de Melo-Diogo D, Gaspar VM, Carvalho MP, Correia IJ (2016) 3D tumor spheroids: an overview on the tools and techniques used for their analysis. Biotechnol Adv 34(8):1427–1441
14. Lazzari G, Couvreur P, Mura S (2017) Multicellular tumor spheroids: a relevant 3D model for the in vitro preclinical investigation of polymer nanomedicines. Polym Chem 8(34): 4947–4969
15. Park SE, Georgescu A, Huh D (1979) Organoids-on-a-chip. Science 364(6444):960–965
16. Ozcelikkale A, Moon H, Linnes M, Han B (2017) In vitro microfluidic models of tumor microenvironment to screen transport of drugs and nanoparticles. Wiley Interdiscip Rev Nanomed Nanobiotechnol 9(5):e1460
17. Prunet A, Lefort S, Delanoë-Ayari H et al (2020) A new agarose-based microsystem to investigate cell response to prolonged confinement. Lab Chip 20:4016–4030
18. Ahmed A, Goodarzi S, Frindel C, Recher G, Riviere C, Rousseau D (2021) Clearing spheroids for 3D fluorescent microscopy combining safe and soft chemicals with deep convolutional neural network. bioRxiv. https://doi.org/10.1101/2021.01.31.428996
19. Riviere C, Prunet A, Fuoco L, Delanoë-Ayari H (2018) Biocompatible hydrogel microwell plate. French Patent, FR1870349A, 27 March 2018
20. Goodarzi S, Prunet A, Rossetti F et al (2021) Quantifying nanotherapeutic penetration using a hydrogel-based microsystem as a new 3D in vitro platform. Lab Chip 21(13): 2495–2510
21. Bregigeon P, Rivière C, Franqueville L, Vollaire C, Marchalot J, Frénéa-Robin M (2022) Integrated platform for culture, observation, and parallelized electroporation of spheroids. Lab Chip 22(13):2489–2501
22. Pluen A, Netti PA, Jain RK, Berk DA (1999) Diffusion of macromolecules in agarose gels: comparison of linear and globular configurations. Biophys J 77(1):542–394
23. Le Duc G, Roux S, Paruta-Tuarez A et al (2014) Advantages of gadolinium based ultrasmall nanoparticles vs molecular gadolinium chelates for radiotherapy guided by MRI for glioma treatment. Cancer Nanotechnol 5(1): 1–14
24. Dunn KW, Kamocka MM, McDonald JH (2011) A practical guide to evaluating colocalization in biological microscopy. Am J Physiol Cell Physiol 300(4):C723–C742

Chapter 15

Selective Targeting of Tumor Cells in a Microfluidic Tumor Model with Multiple Cell Types

Bas van de Crommert, Valentina Palacio-Castañeda, and Wouter P. R. Verdurmen

Abstract

Organ-on-a-chip technology allows researchers to precisely monitor drug efficacy in 3D tissue culture systems that are physiologically more relevant to humans compared to 2D cultures and that allow better control over experimental conditions as compared to animal models. Specifically, the high control over microenvironmental conditions combined with the broad range of direct measurements that can be performed in these systems makes organ-on-a-chip devices a versatile tool to investigate tumor targeting and drug delivery. Here, we describe a detailed protocol for studying the cell-selective targeting of protein drugs to tumor cells on an organ-on-a-chip system using a co-culture consisting of BT-474 cancer cells and C5120 human fibroblasts as an example.

Key words Tumor targeting, Fluorescence microscopy, Protein drug, Organ-on-a-chip, Image analysis

1 Introduction

Organ-on-a-chip technology has the potential to improve many processes during drug discovery and development as it plays a complementary role in bridging outcomes from simple 2D culture systems and animal studies [1]. Particularly in the investigation of larger drugs such as protein or (oligo)nucleotide-based drugs, which due to the species-specificity of their activities have proven to be difficult to study in animal models, this technology has great promise. It can be of tremendous benefit as it can yield physiological outcomes that predict much better what will happen in humans.

Besides physiological relevance, one of the key benefits of organ-on-a-chip systems is that complex biological processes in 3D can be directly monitored with a level of detail that is so far unachievable in animals, due to the use of fluorescent labels or reporting molecules in combination with a variety of readout

Valérie Taly et al. (eds.), *Microfluidics Diagnostics: Methods and Protocols*, Methods in Molecular Biology, vol. 2804, https://doi.org/10.1007/978-1-0716-3850-7_15,

technologies, of which fluorescence microscopy is a prime example. Monitoring can be achieved at the level of the microenvironment, for instance by detecting oxygen levels [2], measuring fluid flow [3], determining mechanical stress [4, 5], or tracking metabolic activity, e.g., through measuring glucose, pH, or reactive oxygen species [6–9].

Next to monitoring environmental factors, cellular fates can be tracked by combining the labeling of fate-associated characteristics, which can be viability, cell death, calcium signaling, or any other characteristic for which a suitable reporter is available, in combination with the labeling of different cell types with unique colors.

Recently, we showed that cell-selective toxicity toward tumor cells through tumor-targeted agents could be investigated on a microfluidic tumor-on-a-chip that contained co-cultures of tumor cells and fibroblasts [10]. By combining microscopic imaging with an image-based analysis using the freely available software Fiji, we were able to demonstrate that the tumor-targeted therapy affected tumor cells, but not fibroblasts. Here, we provide a detailed step-by-step protocol for the execution of experiments of a similar nature, which covers the fabrication of the microfluidic devices, the conditions used for culturing, labeling, and treating the cells, as well as the image-based analysis to investigate cell-selective toxicity of therapeutic agents (Fig. 1).

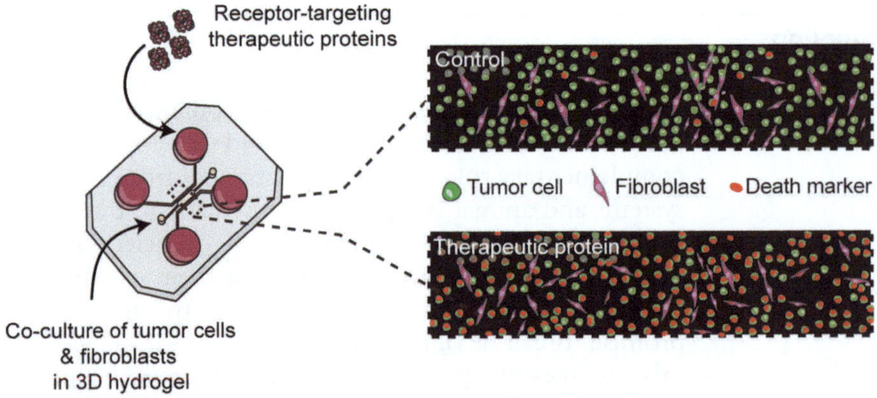

Fig. 1 Method for evaluating selective toxicity in a tumor-on-a-chip system. Schematic representation of the approach detailed in this chapter for the investigation of selective tumor cell targeting using therapeutic proteins. Co-cultures of pre-labeled tumor cells (green) and fibroblasts (magenta) embedded in a 3D hydrogel are initially loaded in the central chamber of the device. The therapeutic proteins are added to the medium reservoirs (5 mm Ø) and diffuse via the side channels to the hydrogel in the central chamber. The effect on cell viability is determined using a death marker (red) that enters dead cells only

2 Materials

For the fabrication of microfluidic devices, a master is required that can be prepared either using photolithography or 3D printing according to procedures that have been discussed in detail elsewhere [11–13]. All buffers should be made with distilled water to ensure sufficient purity.

2.1 Fabrication of Microfluidic Devices

1. Mold: master placed in a casting reservoir (*see* **Note 1**).
2. Polydimethylsiloxane (PDMS) and associated curing agent (*see* **Note 2**).
3. Isopropanol ≥98%.
4. Nitrogen gas (optional) (*see* **Note 3**).
5. Weighing boat or equivalent (*see* **Note 4**).
6. Scalpel.
7. Biopsy punchers (1.5 mm and 5 mm).
8. Single-edge razor blade.
9. Adhesive tape.
10. Precision microscopy slide no. 1.5H (*see* **Note 5**).
11. Analytical scale.
12. Vacuum pump and vacuum chamber.
13. Oven.
14. Plasma cleaner.

2.2 Loading Microfluidic Devices

1. Materials required for standard cell culture maintenance.
2. A cancer cell line, for instance breast cancer cell line MCF7 or BT-474, and primary fibroblasts as representative cell types present in the tumor microenvironment. Cancer cells should be amenable to selective drug targeting.
3. Microfluidic culture medium: medium suitable for both cell lines supplemented with 10% (v/v) fetal calf serum (FCS), 1% (v/v) 100× penicillin-streptomycin (100× stock contains 10,000 U penicillin—10 mg streptomycin per mL) and 2.5 μg/mL amphotericin B (stock solution: 25 mg/mL) (*see* **Note 6**).
4. Dopamine hydrochloride.
5. Tris–HCl buffer: 10 mM Tris–HCl, pH 8.5. Mix and adjust the pH using NaOH. The buffer can be stored for at least 12 months at room temperature.
6. Milli-Q water.
7. Ethanol 70%.

8. Phosphate-buffered saline (PBS): 8 g/L NaCl, 0.2 g/L KCl, 1.38 g/L Na_2HPO_4, 0.26 g/L KH_2PO_4, pH 7.4. Shelf life: 12 months at room temperature.
9. PBS 10×: 80 g/L NaCl, 2 g/L KCl, 13.8 g/L Na_2HPO_4, 2.6 g/L KH_2PO_4, pH 6.5–6.8. Shelf life: 12 months at room temperature.
10. NaOH: 0.5 M.
11. Hydrogel: Rattail collagen type 1 (~11 mg/mL) (*see* **Note 7**).
12. pH strips.
13. Triton ×-100 reagent: 0.1% (v/v) Triton ×-100 in microfluidic culture medium. Prepare fresh every time.
14. Propidium iodide solution: 1 mg/mL propidium iodide in PBS. Shelf life: at least 48 months at −20 °C protected from light.

2.3 Labeling Cells Using Cell Stains

1. Compatible cell stains for the distinct cell types (*see* **Note 8**).
2. Culture medium supplemented with 10% FCS (*see* **Note 9**).
3. High-speed centrifuge.

2.4 Confocal Microscopy Imaging and Image Analysis

1. High-resolution confocal microscopy equipped with a temperature-controlled incubator (*see* **Note 10**).
2. Image analysis software: e.g., freeware Fiji Is Just ImageJ (Fiji) [14].

3 Methods

3.1 Fabrication of Microfluidic Devices

Wear gloves during fabrication steps, clean all objects that come into contact with uncured/cured PDMS with isopropanol prior and after usage and make sure these objects are dry (*see* **Note 11**). It is not necessary to work under sterile conditions during the fabrication.

1. Weigh the PDMS base and curing agent in a weight ratio of 10: 1 (w/w) on a weighing boat and mix well while minimizing the generation of air bubbles (*see* **Note 12**).
2. Pour the mixed PDMS into the clean molds until it reaches a thickness of about 5 mm.
3. Place the molds (without lids) into the vacuum chamber and degas the PDMS until most of the bubbles have disappeared (*see* **Note 13**).
4. Place the molds into the preheated oven and cure them for 2 h at 65 °C.

5. Take the molds out of the oven and remove the PDMS piece containing the microfluidic design from the molds by cutting along the edge of the casting reservoir using a scalpel. Punch holes for the medium reservoirs (5 mm) and inlets for the main compartment (1.5 mm) from the side of the design using biopsy punchers. Cut overhangs using the single-edge razor blade to make it fit the microscope slide and cover the microfluidic design with adhesive tape (*see* **Note 14**).
6. Remove the tape from the microfluidic device and place it with the design facing upward together with the clean objective slide on the sample holder and place it in the sample chamber of the plasma cleaner.
7. Treat the samples using the plasma cleaner according to manufacturer's instructions.
8. Take out the sample holder and gently press the treated microfluidic device (with the design facing the objective slide) on the objective slide and make sure the whole surface of the PDMS is in contact with the functionalized glass slide. Cover the medium reservoirs and inlets with adhesive tape to prevent contaminants from entering the channels. Wait for at least 24 h before proceeding with the polydopamine treatment and subsequent loading of the microfluidic devices to allow hydrophobic recovery of the objective slide (*see* **Note 15**).

3.2 Loading Microfluidic Devices

It is highly recommended to keep the microfluidic devices in Petri dishes after removal of the adhesive tape to minimize the risk of contaminants entering the channels. It is not necessary to work under sterile conditions for **steps 1–3**. Starting from **step 4**, the sample working conditions should be as sterile as possible. All media added to the cells during loading and treatment should be preheated in a water bath set at 37 °C.

1. Prior to loading the microfluidic devices, freshly prepare a 2 mg/mL dopamine hydrochloride solution in 10 mM Tris–HCl pH 8.5 and slowly pipet the solution in the main channel. Incubate the microfluidic devices for 1 h at room temperature protected from light (*see* **Note 16**).
2. Remove the dopamine hydrochloride solution and wash the main channels twice with Milli-Q, followed by once with 70% ethanol. Remove the 70% ethanol solution.
3. Place the microfluidic devices in the oven at 65 °C for 2 h (*see* **Note 17**).
4. Just before the 2 h have passed, detach, count, and pellet the required volume and concentration of cells (20×10^6 cells/mL in hydrogel-cell suspension) by centrifugation (*see* **Note 18**).

5. Label the cells according to the manufacturer's protocol for the cytoplasmic dye that is being used (e.g. CellTrace or equivalent).
6. Take the microfluidic devices out of the oven and let them cool down before loading them in **step 8**.
7. Prepare the hydrogel-cell suspension on ice (*see* **Note 19**):
   For collagen type 1 (100 μL hydrogel-cell suspension): Carefully mix in this order: 7 μL 10× PBS, a volume of Milli-Q enough to reach a final hydrogel-cell suspension volume of 100 μL, NaOH 0.5 M to reach a pH of ~7.4 (typically around 1–2 μL), collagen type 1 to a final concentration of 4 mg/mL, and 30 μL cell suspension in culture medium. Mix gently to minimize the introduction of air bubbles. It is recommended to measure the pH before adding the cells using pH strips to confirm that the pH is physiological (pH 7.0–7.5).
8. Place the chips in new sterile Petri dishes and slowly pipet the hydrogel-cell suspension into the inlet of the main compartment. By gently and slowly loading the hydrogel-cell suspension, you will ensure that the gel remains in the main compartment and does not spread into the side channels.
9. Place the microfluidic devices in the incubator at 37 °C for 30 min to allow polymerization of the hydrogel.
10. After polymerization, add 100 μL of culture medium to each of the four medium reservoirs making sure that medium fills up the side channels (*see* **Note 20**). Add 1 mL medium on top of the microfluidic device to counter evaporation.
11. Add +/− 5 mL PBS to each Petri dish to limit evaporation of the medium.
12. Carefully place the microfluidic devices in the incubator at 37 ° C and allow the cells to grow for 24 h before starting treatment.
13. When starting treatment, prepare your incubation medium by diluting the drug of interest in microfluidic culture medium. Remove the medium present on the microfluidic device and in the medium reservoirs and replace it with the same volume of incubation medium. Replace the (incubation) medium every 2 days to ensure cells have sufficient nutrients available.
14. After treatment and prior to imaging, dilute the 1 mg/mL propidium iodide stock solution to 20 μg/mL in microfluidic culture medium. Add 100 μL of the 20 μg/mL propidium iodide solution to each one of the medium reservoirs and incubate for 1 h at 37 °C. For the positive control (all dead cells), 0.1% (v/v) Triton ×-100 is added to the 20 μg/mL propidium iodide solution and co-incubated for 1 h at 37 °C (Fig. 1).
15. Proceed with imaging.

3.3 Confocal Microscopy Imaging

Set the correct excitation and emission wavelengths for the Cell-Trace reagents and propidium iodide. During confocal microscopy, it is not necessary to work under sterile conditions. Place and keep the Petri dishes with the live cells in the microfluidic devices into the incubator at 37 °C.

1. Start the software on the confocal microscope. If the spectra overlap with each other the scanning should be done sequentially. Switch on the brightfield for orientation of your image acquisition in the microfluidic device. Use a minimally sufficient laser power to reduce photobleaching of the signal (*see* **Note 21**).
2. Place the microfluidic device on the stage and focus on the lowest plane containing fluorescent signal and set this as the bottom of the Z-stack. Move the z direction to the highest plane containing cells and fluorescent signal and set this as the top of the Z-stack (*see* **Note 22**).
3. Select the desired resolution (minimally 1024 × 1024 pixels) and set the imaging slice size of the Z-stack to 10 microns (*see* **Note 23**).
4. Start the imaging of the microfluidic devices and check before proceeding with other conditions if the set z-slice positions still capture the cells in focus.
5. Repeat steps for all other conditions.

3.4 Image Analysis

The steps for image analysis described below are designed for working with the freely available image-processing software *Fiji* and correspond to the steps in Fig. 2. It is advised to create a macro in *Fiji* to facilitate the image analysis of many different conditions. The macro used for image analysis can be found in Table 1.

1. Open *Fiji* and load a compatible data file obtained from the microscope. For each condition, select the z-slice of interest and isolate this slice from the stack. Duplicate the isolated z-slice and work further with the duplicate.
2. Separate the channels from the composite image into single channels.
3. Apply a median filter to all channels of the z-slice (*see* **Note 24**).
4. Set the threshold for each channel in a way that only the cell or propidium iodide area is shown as positive signal. Apply the same threshold for the same channel between samples that you will directly compare.
5. Create a selection of the positive area corresponding to the cells or propidium iodide and convert this selection into a region of interest (ROI). Add the ROI of all channels to the ROI

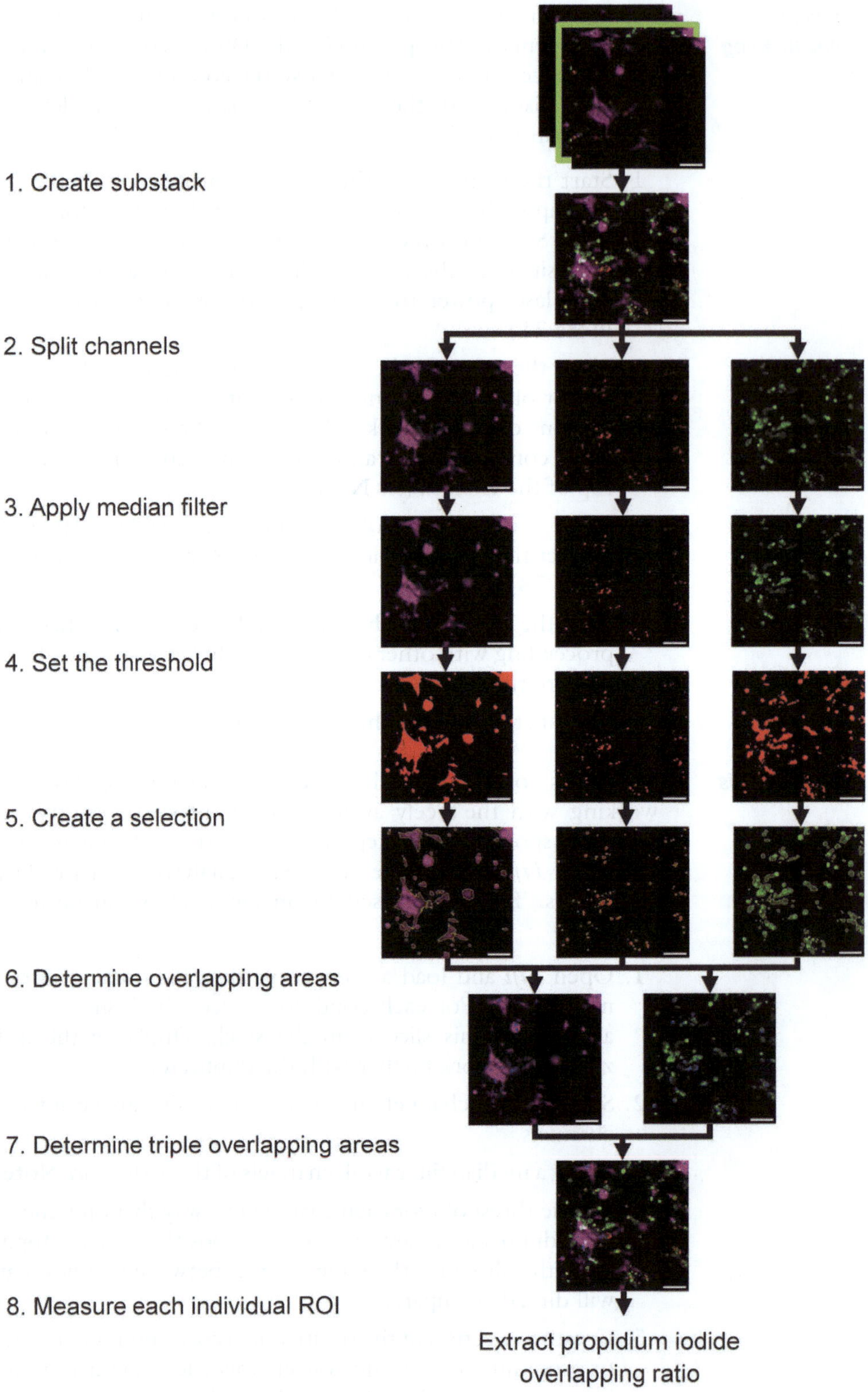

Fig. 2 Flow scheme visualizing the image analysis steps from raw data input to quantitative output of the microscopy data. Further explanations about individual steps can be found in Subheading 3.4 Image analysis.

manager and rename them to the corresponding channel. Confirm that in the isolated original z-slice, the ROIs adequately match the cells or propidium iodide area.

6. Select the ROI of one of the cell channels and the ROI of the propidium iodide channel in the ROI manager, click "more" and select "AND." This creates a new selection of the overlapping ROIs of the cell and propidium iodide. Add this selection to the ROI manager and repeat this step for the other cell channels.
7. Select the ROIs of all three channels, create a new selection of the triple overlapping areas and add this selection to the ROI manager (*see* **Note 25**).
8. Measure the area and display the label of the measurement. Select and measure each individual ROI in the ROI manager.
9. Subtract the triple overlapping area from the cell, propidium iodide, and overlapping areas.
10. Divide the overlapping area by the cell area for all cell lines and plot the obtained ratios in a column chart (Fig. 3).

4 Notes

1. We typically use a master created by photolithography on a silicon wafer with a diameter of 5 cm. This master is placed in a Petri dish or aluminum dish with a diameter of 6 cm. Casting reservoirs should ideally be selected to fit the masters well, as this reduces PDMS waste. Casting reservoirs with a height of at least 1 cm above the master is preferred to prevent spillage of PDMS during bubble removal with vacuum.
2. PDMS is a silicon-based elastomer that is widely used for the fabrication of microfluidic devices due to its biocompatibility, elasticity, optical transparency, and low autofluorescence [15].
3. Nitrogen gas or compressed air can be used to remove any residual isopropanol left in the microstructures of the master during cleaning.

Fig. 2 (continued) For the representative analysis, we used images that were acquired using an experimental condition that we published before [10]. Briefly, cells were treated with 10 nM Ec1-ETA, a fusion protein toxin between a designed ankyrin repeat protein (DARPin) that binds with high affinity to epithelial cell adhesion molecule (EpCAM), which is often highly overexpressed on tumor cells, and ETA, which refers to the translocation and catalytic domain of the *Pseudomonas aeruginosa* exotoxin A (ETA). A tumor cell-selective delivery of these proteins halts protein synthesis specifically in tumor cells, which is followed by tumor cell death. Scale bars represent 100 μm. PI, propidium iodide

Table 1
Fiji macro for image analysis

1. Create substack.
```
selectWindow("Example");
run("Make Substack...", "channels=1,2,3 slices=6");
```

2. Split channels and create duplicates.
```
run("Split Channels");
selectWindow("C5120 channel");
selectWindow("Propidium iodide channel");
run("Duplicate...", " ");
selectWindow("C5120 channel");
run("Duplicate...", " ");
selectWindow("BT-474 channel");
run("Duplicate...", " ");
```

3. Apply median filter.
```
selectWindow("Propidium iodide channel-1");
run("Median...", "radius=2");
selectWindow("C5120 channel-1");
run("Median...", "radius=2");
selectWindow("BT-474 channel-1");
run("Median...", "radius=2");
```

4. Set the threshold.
```
selectWindow("Propidium iodide channel-1");
setAutoThreshold("Default dark");
run("Threshold...");
selectWindow("C5120 channel-1");
setThreshold(20, 255, "raw");
selectWindow("Propidium iodide channel-1");
setThreshold(20, 255, "raw");
selectWindow("BT-474 channel-1");
setThreshold(55, 255, "raw");
```

5. Create a selection and convert into ROIs.
```
selectWindow("Propidium iodide channel-1");
run("Create Selection");
roiManager("Add");
selectWindow("C5120 channel-1");
run("Create Selection");
roiManager("Add");
selectWindow("BT-474 channel-1");
run("Create Selection");
roiManager("Add");
roiManager("Select", 0);
roiManager("Rename", "Propidium iodide channel");
roiManager("Select", 1);
roiManager("Rename", "C5120 channel");
```

(continued)

Table 1
(continued)

```
    roiManager("Select", 2);
    roiManager("Rename", "BT-474 channel");
6.  Determine overlapping areas.
    roiManager("Select", newArray(0,1));
    roiManager("AND");
    roiManager("Add");
    roiManager("Select", 3);
    roiManager("Rename", "C5120:Propidium iodide overlap");
    roiManager("Select", newArray(0,2));
    roiManager("AND");
    roiManager("Add");
    roiManager("Select", 4);
    roiManager("Rename", "BT-474:Propidium iodide overlap");
7.  Determine triple overlapping areas.
    roiManager("Select", newArray(0,1,2));
    roiManager("AND");
    roiManager("Add");
    roiManager("Select", 5);
    roiManager("Rename", "Triple overlap");
8.  Set measurements and measure each individual ROI.
    run("Set Measurements...", "area display redirect=None decimal=3");
    roiManager("Select", 0);
    roiManager("Measure");
    roiManager("Select", 1);
    roiManager("Measure");
    roiManager("Select", 2);
    roiManager("Measure");
    roiManager("Select", 3);
    roiManager("Measure");
    roiManager("Select", 4);
    roiManager("Measure");
    roiManager("Select", 5);
    roiManager("Measure");
```

4. A falcon tube can be used for weighing and mixing instead of a weighing boat, since it can be centrifuged (3 min at 300 × *g*) after mixing the PDMS in it to remove most of the air bubbles. This saves time during bubble removal under vacuum from the microfluidic devices. Check after centrifugation if there is one homogenous layer, otherwise repeat the mixing and centrifugation.

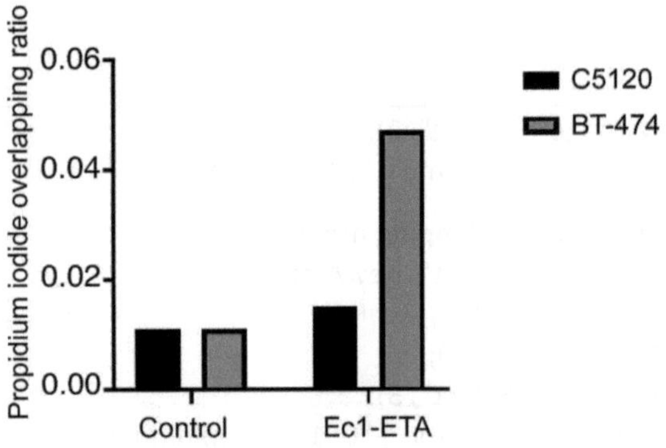

Fig. 3 Quantification of the overlap between propidium iodide and the cell channels. Quantification for the example data set was done on only one slice per condition. The control samples correspond to the untreated samples. For sample details see Fig. 2 and the reference therein

5. Check beforehand the working distance of the objective lens on the microscope to select the compatible thickness of the microscope slide. It is highly recommended to use high-precision microscope slides because they have low variation in thickness.
6. Microfluidic devices are often placed in a Petri dish while culturing. Adding and removing solutions is associated with a risk of bacterial or fungal contamination. While we recommend working under sterile conditions, we include antibiotics and antimycotics as a further precaution for long-term experiments.
7. Different commercially available hydrogels can be selected based on the tissue type to be mimicked. We prefer to work with rattail collagen type I or Matrigel. Many synthetic hydrogels with a wide variety of properties are nowadays available.
8. Different commercially available cell stains can be selected to meet experimental requirements (e.g., CellTrace or equivalent). The cytoplasm of cells can be labeled by covalently coupling fluorophores to available functional groups, usually primary amines. Cell labeling allows distinguishing between cell types and enables cell type-selective quantification approaches during image analysis.
9. The unbound cell stain reagent can diffuse into the medium where it can react with primary amine groups of serum proteins. Following several washes, this removes any unbound cell stain reagent that could induce background fluorescence.
10. The temperature-controlled microscope incubator is optional if the duration of sample exposure to room temperature is limited.
11. It is recommended to use gloves throughout the fabrication procedure. PDMS is classified as a nonhazardous substance, but because of its adhesive properties, it is difficult to remove

from surfaces if spilled. Gloves also make sure that no fingerprints or oils from the skin are left behind on the microfluidic devices that would create artifacts during imaging. Cleaning all surfaces that come into contact with uncured PDMS with isopropanol ensures that no contaminants get cured into the PDMS. We use isopropanol because of its fast evaporation leaving minimal residual substances behind, being compatible with viable cell cultures afterwards. Nitrogen gas can be used to speed up evaporation.

12. The required amount of PDMS can be determined beforehand by adding Milli-Q water to the molds (+/− 5 mm above master) and using this volume as reference. Mixed PDMS has a specific gravity of 1.03 at 25 °C so the same volume/weight of PDMS can be used as determined using Milli-Q water. Cover the scale and the pouring workplace with paper towels. This makes cleaning spillage much easier. For mixing, a stainless-steel spatula can be used.

13. Evacuate the chamber in multiple cycles until most of the air bubbles have disappeared. Make sure that no air bubbles are left at the bottom near the microfluidic design. Some air bubbles still present at the surface will diffuse in the oven.

14. When working with masters created by photolithography, extra caution is required not to apply too much pressure while cutting out the PDMS to prevent breaking the master. Avoid touching the microfluidic design since this can collapse the small structures. When cutting away the overhang, it is advised to cut away the four corners of the rectangle to create an octagon to remove irregularities from the edges and improve bonding to the microscope slide.

15. When bonding the PDMS chips to the microscope slide press gently around the design to avoid possible collapse of small structures in the design. Oxygen plasma treatment increases the hydrophilicity of the treated surface by introducing polar functional groups. Hydrophobic recovery decreases surface wettability and thereby ensures that during the next step, the dopamine solution does not spread into the medium side channels. To speed up the hydrophobic recovery, the chips can be placed for ~2 h in the oven at 65 °C.

16. Place the tube with residual dopamine hydrochloride solution under the same conditions as the microfluidic devices and check if the solution turns dark after 1 h to confirm polymerization of the dopamine. Dopamine treatment functions as surface modification of the microscope slide and PDMS improves collagen binding.

17. Placing the microfluidic devices in the oven ensures complete evaporation of the ethanol and elimination of any bacteria present.
18. When working with co-cultures the number of cells per cell line is typically equal in our experiments, though this can be adjusted based upon experimental needs. We generally aim for a final seeding density of 20×10^6 cells/mL in collagen-cell suspension.
19. It is important to keep all the components of the collagen solution on ice. This prevents a premature polymerization. Avoid touching the bottom of the Eppendorf tubes containing collagen because the heat of your skin can easily be transferred to the Eppendorf tube and induce polymerization.
20. When adding the medium to the reservoirs, it is important to fill the medium channels fully so that medium diffusion occurs unhindered. This can be done in two ways: either by filling one medium reservoir and by creating a negative pressure through the channel from the empty reservoir or by creating a positive pressure through the channel from one of the empty reservoirs with a filled pipet tip. To create this negative/positive pressure, a P200 pipet can be used. Gently align the (filled) tip of the pipet with the medium channels and aspirate/dispense the medium through the channels.
21. When working with multiple fluorophores, it is always advised to check the spectrum compatibility beforehand to ensure that no bleed-through between channels occurs. If there is a risk for bleed-through between channels, use a sequential imaging approach. Be aware that high-intensity excitation and prolonged imaging increase the photobleaching of fluorophores.
22. Placing a wet paper tissue close to the microfluidic device during imaging limits the evaporation of the liquid in the microfluidic device.
23. The slice size is set to 10 microns, which approximates the average diameter of a human cell. This slice thickness therefore ensures that few cells will be missed with each new z-slice, and it minimizes the total amount of z-slices.
24. Applying a median filter reduces background noise. This neglects very high local signal intensities due to background noise and thus decreases the amount of false positive pixels when setting the threshold, thereby enhancing the accuracy of cell area determination.
25. The triple overlapping area is the propidium iodide overlapping signal that cannot be linked to one single cell type. By subtracting the triple overlapping area from the cell, propidium iodide, and overlapping areas, the signal that is non-cell type specific is excluded from analysis.

References

1. Low LA, Mummery C, Berridge BR, Austin CP, Tagle DA (2021) Organs-on-chips: into the next decade. Nat Rev Drug Discov 20(5): 345–361
2. Palacio-Castañeda V, Velthuijs N, Le Gac S, Verdurmen WPR (2022) Oxygen control: the often overlooked but essential piece to create better in vitro systems. Lab Chip 22(6): 1068–1092
3. Stone HA, Stroock AD, Ajdari A (2004) Engineering flows in small devices: microfluidics toward a lab-on-a-chip. Annu Rev Fluid Mech 36:381–411
4. Kaarj K, Yoon JY (2019) Methods of delivering mechanical stimuli to organ-on-a-chip. Micromachines (Basel) 10(10):700
5. Oleaga C, Lavado A, Riu A, Rothemund S, Carmona-Moran CA, Persaud K, Yurko A, Lear J, Narasimhan NS, Long CJ, Sommerhage F, Bridges LR, Cai YQ, Martin C, Schnepper MT, Goswami A, Note R, Langer J, Teissier S, Cotovio J, Hickman JJ (2019) Long-term electrical and mechanical function monitoring of a human-on-a-chip system. Adv Funct Mater 29(8): 1805792
6. Lo JF, Sinkala E, Eddington DT (2010) Oxygen gradients for open well cellular cultures via microfluidic substrates. Lab Chip 10(18): 2394–2401
7. Lo JF, Wang Y, Blake A, Yu G, Harvat TA, Jeon H, Oberholzer J, Eddington DT (2012) Islet preconditioning via multimodal microfluidic modulation of intermittent hypoxia. Anal Chem 84(4):1987–1993
8. Tanumihardja E, Slaats RH, van der Meer AD, Passier R, Olthuis W, van den Berg A (2021) Measuring both pH and O2 with a single on-chip sensor in cultures of human pluripotent stem cell-derived cardiomyocytes to track induced changes in cellular metabolism. ACS Sens 6(1):267–274
9. Palacio-Castañeda V, Kooijman L, Venzac B, Verdurmen WPR, Le Gac S (2020) Metabolic switching of tumor cells under hypoxic conditions in a tumor-on-a-chip model. Micromachines (Basel) 11(4):382
10. Palacio-Castañeda V, Dumas S, Albrecht P, Wijgers TJ, Descroix S, Verdurmen WPR (2021) A hybrid in Silico and tumor-on-a-chip approach to model targeted protein behavior in 3D microenvironments. Cancers (Basel) 13(10):2461
11. Sackmann EK, Fulton AL, Beebe DJ (2014) The present and future role of microfluidics in biomedical research. Nature 507(7491): 181–189
12. Miri AK, Mostafavi E, Khorsandi D, Hu SK, Malpica M, Khademhosseini A (2019) Bioprinters for organs-on-chips. Biofabrication 11(4): 042002
13. Sun H, Jia Y, Dong H, Dong DB, Zheng JP (2020) Combining additive manufacturing with microfluidics: an emerging method for developing novel organs-on-chips. Curr Opin Chem Eng 28:1–9
14. Schindelin J, Arganda-Carreras I, Frise E, Kaynig V, Longair M, Pietzsch T, Preibisch S, Rueden C, Saalfeld S, Schmid B, Tinevez JY, White DJ, Hartenstein V, Eliceiri K, Tomancak P, Cardona A (2012) Fiji: an open-source platform for biological-image analysis. Nat Methods 9(7):676–682
15. Borok A, Laboda K, Bonyar A (2021) PDMS bonding technologies for microfluidic applications: a review. Biosensors (Basel) 11(8):292

Index

Valérie Taly et al. (eds.), *Microfluidics Diagnostics: Methods and Protocols*, Methods in Molecular Biology, vol. 2804, https://doi.org/10.1007/978-1-0716-3850-7,

Q

R

S

T

MIX
Papier aus verantwortungsvollen Quellen
Paper from responsible sources
FSC® C105338

If you have any concerns about our products,
you can contact us on
ProductSafety@springernature.com

In case Publisher is established outside the EU,
the EU authorized representative is:
Springer Nature Customer Service Center GmbH
Europaplatz 3, 69115 Heidelberg, Germany

Printed by Libri Plureos GmbH
in Hamburg, Germany